NATIVE INSECTS OF AOTEAROA

NATIVE INSECTS
OF AOTEAROA

Julia Kasper and Phil Sirvid

TE PAPA

PRESS

CONTENTS

INTRODUCTION

Everybody has seen insects, but very few give them much thought. For just a moment, we now invite you to give insects no thought at all and try to imagine a world without them. While it might be a pleasure never to see a fly buzzing around the living room again, within a very short time the cruel reality of what an absence of insects means will become apparent. The loss of bees and other pollinators would have devastating effects on agriculture and horticulture. And although we might not find a maggot-ridden animal corpse delightful, how much worse would it be if there were nothing to help break down carcasses? Think, too, of all the other plants and animals that depend on insects for their survival; whole food chains would collapse. And these are just a few examples.

As notable entomologist EO Wilson famously said, insects are the little things that run the world. They represent a myriad of marvels in miniature, encompassing a rich array of forms and every conceivable colour. To the naked eye, insects often appear to be small and indistinct, with unpredictable movements. Put an insect under the microscope, however, and its complexity soon becomes apparent. Patient observation in the field or in the laboratory will often reveal a rich repertoire of behaviour.

In terms of diversity, insects are inarguably the most successful group on the planet – they account for over half the known species of multi-cellular organisms on Earth. Globally about a million species of insect have been described and named, but the true number of species may be as much as five to ten times greater. Even for those insects that are already part of the scientific record, for the most part we have only the barest idea of what they do and how they live their lives. Indeed, sometimes all we know of a species is based on a specimen or two collected decades ago. However, some insects are exceedingly well-studied. For example, we owe much of our modern understanding of genetics and developmental biology to research on a species of small vinegar fly, *Drosophila melanogaster*.

Insects can be found in every conceivable terrestrial or freshwater habitat, from caves to mountain-tops, from deserts to lakes and streams, from grasslands to forest, from Antarctica to geothermal hot pools, and even in our homes and gardens.

This book covers a range of Aotearoa New Zealand's insects, but not the allied groups of animals that might also be colloquially called bugs or creepy crawlies. Insects are not the only animals that have six legs. The class Entognatha, for example, are six-legged arthropods that are separated from insects by their ability to retract their mouthparts into their heads. They are not covered in this volume, and nor are other kinds of arthropod such as spiders and centipedes.

THE INSECT FAUNA OF AOTEAROA

The insect fauna of Aotearoa New Zealand has been described as disharmonic, meaning that we have few or no examples of some groups that might be expected to be found here, whereas others are surprisingly diverse. For example, we have only one species of mantis and our endemic ant fauna sits at a paltry eleven species. In contrast, Australia has nearly 120 mantis species and the great majority of its nearly 1300 known ant species are unique to that country. Our butterfly fauna is also small, consisting of around two dozen species, some of which are only occasional visitors blown across the Tasman Sea. The United Kingdom, which has a similar land area to Aotearoa but less unmodified natural habitat, has around sixty species. In contrast to butterflies, some moth families are very diverse. With about 250 species, the family Oecophoridae makes up about 12.5 percent of our moth fauna.

We also have a mix of old lineages that have persisted during the long isolation of the Aotearoa landmass since the break-up of Gondwana, and more-recent lineages descended from species that have colonised Aotearoa since then. As travel over large oceanic gaps like the Tasman Sea is difficult for insects that depend on fresh water for their survival, our aquatic insects include examples of lineages that have always been present here. In contrast, molecular data shows that our entire cicada fauna is descended from colonists from New Caledonia and Australia. Some groups, such as the beetle family Zopheridae, are a mix of old and new, with lineages that seem to have persisted since the Cretaceous alongside others descended from more recent arrivals.

One peculiarity of our insect fauna is the number of flightless species. To take one example, it is estimated that about a quarter of our crane fly species don't have wings. Flightlessness in Aotearoa is classically explained by an absence of mammalian predators, such that many insect species were ill-equipped to cope with their later arrival. For example, some species of giant wētā once present on the mainland have been reduced to populations on islands free of introduced predators such as rats.

Aotearoa is also home to a range of alpine insects, including butterflies, moths, grasshoppers, stoneflies, cicadas and wētā. Some of these insects have behavioural or physiological adaptations for survival in sometimes sub-zero temperatures. The mountain stone wētā (*Hemideina maori*) can freeze solid overnight, then thaw out and function normally the next day; it also has the distinction of being the world's largest freezing-tolerant insect. Other species, such as the black mountain ringlet butterfly, bask in the sun on warm stones, using their dark wings to capture more sunlight. At the other end of the scale, there are fly larvae and beetles living in geothermal hot pools at temperatures over 45°C.

The most recent general inventory of Aotearoa insects, published in 2010, lists over 12,000 described species and subspecies, more than 90 percent of them being unique to Aotearoa. It also estimated our insect fauna to be over 20,000 species in total, although at the present rate of discovery it would take centuries to describe them all. Most of our moth species have been described and named, although many new species are still being discovered. Other orders are less well-studied, including major orders such as Diptera, which may have over 1500 undescribed species. And species descriptions are just the start; there is so much more to be learned from them than that.

INSECT COLLECTIONS AT TE PAPA

Founded in 1865 as part of the Colonial Museum, Te Papa's entomology collection now has more than a million specimens. As well as insects, the collection includes allied groups such as arachnids (e.g. spiders, mites, harvestmen, false scorpions), myriapods (centipedes and millipedes), Entognatha (e.g. springtails) and tardigrades (water bears). Specimens are stored in one of three ways. Insects that are sufficiently large and durable are mounted on pins and stored in drawers in cabinets. Softer-bodied specimens that would shrivel or potentially rot if left exposed to air are usually preserved in vials of 70 percent ethanol. Extremely small specimens are often mounted permanently on glass slides for microscopic examination.

The entomology collection covers all insect orders found in Aotearoa, and is particularly strong in beetles, cicadas, wētā, springtails, moths, lice, fleas and spiders. There are also substantial foreign collections. Important collectors include GV Hudson (the source of many of the illustrations in this book), JT Salmon, RLC Pilgrim, RR Forster and Sir Charles Fleming.

When a species is described for the first time, a specimen is selected by the author to act as the representative example of that species. This is called a holotype, and the entomology collection holds nearly 1200 of these. If there is any doubt about the identity of a specimen, the holotype specimen can be thought of as the ultimate reference specimen.

The entomology collection is a valuable record of biodiversity and of the presence of a species at a given place and time. It is an important scientific resource; experts from all over the world visit it or borrow specimens to study. Scientists are not the only users of the collection – it has also been a source of inspiration for artists, photographers, and even advertising agencies and major movie-makers. Specimens from the collection also feature in Te Papa exhibitions such as *Te Taiao Nature* and *Bug Lab*.

Phil Sirvid, Julia Kasper

CLASSIFYING AND IDENTIFYING INSECTS

INSECT CLASSIFICATION – A BEGINNER'S GUIDE

To make sense of the sheer abundance of life, biologists use a ranked classification system, with each step down in rank becoming more specific as the members of that rank have more and more characteristics in common with each other than with anything else. Kingdom is the broadest level, then phylum, class, order, family, genus – and unsurprisingly, it ends with species! Taxonomists might use finer or broader divisions between these ranks (for example we can have a subphylum or a superorder), but the major ranks usually suffice for most purposes.

Taking the Wellington tree wētā as an example, it's clearly an animal and so, like birds, spiders or fish, it belongs to Kingdom Animalia. However, it is a very different kind of animal to a bird or a fish, so we group these in different phyla. Birds and fish are animals with backbones, and so belong to the phylum Chordata; wētā and spiders have features like an exoskeleton and jointed legs in common, so are grouped into the phylum Arthropoda.

Insects and spiders are similar enough to be grouped together as arthropods, however, they also have some differences, such as the number of legs. This allows us to separate them by putting wētā into class Insecta and spiders into class Arachnida.

Now we are down to order level. This is where we meet some familiar (and not so familiar) groupings of similar kinds of insect – such as moths and butterflies (order Lepidoptera), beetles (order Coleoptera) and bees, ants and wasps (order Hymenoptera). Our wētā might have the same number of legs as beetles or wasps, but they have much more in common with grasshoppers, locusts and crickets, so are members of the order Orthoptera.

Following the same rule of grouping similar things together, we then go down through family, genus and species levels. The Wellington tree wētā shares more anatomical features with giant wētā and other tree wētā than it does with grasshoppers, so these wētā reside in the family Anostostomatidae. However, giant wētā and tree wētā are distinct from each other; they are placed in the genera *Deinacrida* and *Hemideina* respectively. As the Wellington wētā has features that set it apart from other tree wētā, it gets its own species name, which always appears together with the genus name. Thus, our wētā is *Hemideina crassidens*, which belongs to kingdom Animalia, phylum Arthropoda, class Insecta, order Orthoptera and family Anostostomatidae.

Below is a full list of Aotearoa New Zealand insect orders. Asterisks (*) denote orders not covered by species entries in this book, see page 16 for further explanation.

Archaeognatha (bristletails)*
Zygentoma (silverfish)*
Odonata (dragonflies and damselflies)
Ephemeroptera (mayflies)
Plecoptera (stoneflies)
Dermaptera (earwigs)
Blattodea (cockroaches and termites)
Mantodea (mantids)
Phasmatodea (stick insects)
Orthoptera (crickets, grasshoppers, wētā, katydids and relatives)
Hemiptera (true bugs)
Psocodea (psocids, booklice, barklice and parasitic lice)
Thysanoptera (thrips)
Hymenoptera (bees, wasps and ants)
Megaloptera (dobsonflies)
Neuroptera (lacewings)
Strepsiptera (twisted wing parasites)*
Coleoptera (beetles)
Mecoptera (scorpionflies)*
Diptera (flies)
Siphonaptera (fleas)
Trichoptera (caddisflies)
Lepidoptera (butterflies and moths)

BASIC INSECT ANATOMY

Insects have a three-part body structure consisting of a head, a thorax and an abdomen. They have an exoskeleton and must shed their skins periodically in order to grow (see the section on insect lifecycles on page 15). The eyes, antennae and mouthparts are located on the head.

Most insects have two compound eyes made up of an array of photoreceptors. These don't provide crisp, detailed images the way human eyes do, but they are extremely good at detecting movement. Many insects may also have simple eyes called ocelli; these are much better than compound eyes at detecting changes in light and may also have a role in flight stability. Other functions, including image perception and regulation of circadian rhythms, may also be possible. Insect eyes use a different part of the spectrum than humans. We see more of the red part of the spectrum, whereas insects see more of the ultraviolet portion. Flowers can look very different under UV light, and this can help attract pollinating insects.

Insect antennae are commonly called feelers, but have more than a tactile role. They carry many other sensors that can pick up scents, register hot or cold, or perceive air currents, and are particularly important for the detection of pheromones. This is most obviously seen in male moths, which commonly have much larger, feather-like antennae than females. This greatly increases the number of sensors and helps the males find females despite the infinitesimally small concentrations of pheromones in the air.

Insect mouthparts are very diverse, covering a range of feeding types including biting and chewing, piercing and sucking, and sponging. The type of mouthparts is often consistent across an insect order, although the exact form may vary greatly, sometimes even within the same species. For example, grasshoppers and tree wētā are both members of the order Orthoptera and have chewing mouthparts, but in tree wētā females these take the form of small, sharp mandibles for enlarging cavities in trees, while males have much larger mandibles that are used to defend access to the females with which they cohabit.

The thorax is located between the head and the abdomen on the insect body; it is where the legs and wings are located. Insects are the only invertebrates that have self-powered flight. There are two main flight systems. The first is called direct flight, as the wing bases are directly attached to the flight muscles. Contracting one set of flight muscles raises the wings, while contracting a second set lowers them. This system can also allow for independent control of each wing, permitting extremely fine manoeuvring, most famously seen in the acrobatic flight of dragonflies. The majority of flying insects, however, use an indirect flight muscle system in which both the wing bases and the flight muscles are attached to the cuticle of the insect instead of to each other. Contraction and relaxation of the flight muscles change the shape of the cuticle, which in turn moves the wings.

The abdomen is the rearmost portion of the insect body and is where the organs for respiration, excretion, digestion and reproduction are located.

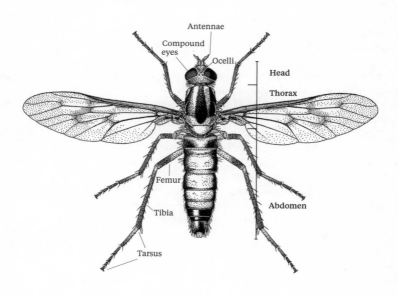

INSECT LIFECYCLES

Insects generally start life with an egg stage, although some, such as aphids, can birth live young. Insects are arthropods and thus have exoskeletons. This can be likened to growing inside a suit of armour; eventually the insect must shed the old exoskeleton to grow, in a process called ecdysis or moulting. This typically involves a split forming along the back of the old exoskeleton, and the next stage (or instar) pulling itself clear through this opening. At this point the insect is very soft, and it inflates itself to a slightly larger size before the new exoskeleton hardens. Moulting is a time of extreme vulnerability as the insect is defenceless and immobile until the exoskeleton is sufficiently hard.

Insects have one of two lifecycle types: they are either holometabolous or hemimetabolous. Holometabolous insects have a pupal stage during which they undergo complete metamorphosis. Immature stages can look extremely different to the final adult stage (e.g. caterpillars and butterflies). The pupal stage doesn't feed and is usually immobile (mosquito pupae are a well-known exception). Inside the pupa, the larval structures break down and adult structures (including any wings) form.

Hemimetabolous insects undergo partial metamorphosis and lack a pupal stage. Compared with holometabolous insects, changes between life stages are more gradual and immature stages resemble the adults (e.g. wētā) unless the immature stages live in a very different environment than the adults (e.g. dragonfly nymphs). If a hemimetabolous insect has wings, in almost all cases these do not fully form until the insect has reached the final, adult stage of its life. Mayflies (order Ephemeroptera) are the exception. They have a winged but sexually immature subimago stage that undergoes an additional moult before the reproductively capable adult stage emerges.

All members of an order have the same lifecycle type. For example, all Lepidoptera (butterflies and moths) are holometabolous, while all Odonata (dragonflies and damselflies) are hemimetabolous. Logically, if unsurprisingly, the holometabolous and hemimetabolous orders are grouped into the superorders Holometabola and Hemimetabola, respectively.

ABOUT THIS BOOK

As with the other books in this series, we were tasked with choosing images to depict the animals we wanted to write about. Many species descriptions only have illustrations of key features rather than whole animals, and some orders – usually small or obscure orders such as scorpionflies (Mecoptera) – do not always have suitable illustrations. For that reason, some orders are not represented in this book.

The illustrations included in the book come from three key sources. In 1992, Des Helmore began work as an entomological illustrator for the Entomology Division of the Department of Scientific and Industrial Research (DSIR). After the DSIR was restructured into Crown institutes, Helmore continued this work for the New Zealand Arthropod Collection, New Zealand's largest entomology collection and part of Manaaki Whenua – Landcare Research. He retired in 2006 and produced over 1000 illustrations, mostly for the *Fauna of New Zealand* series. These are now available on Wikimedia Commons.

Helmore's illustration technique utilised a stereomicroscope with a drawing tube. This allows an image to be 'seen' on the paper, thus enabling the exact proportions of a specimen to be traced. Helmore's drawings were typically made using Indian ink and technical drawing pens. The illustration was double or triple publication size, making an accurate portrayal easier, while imperfections were minimised in the smaller, published version.

Amateur entomologist George Vernon Hudson's specimen collection and scientific contribution is well recognised, but he also leaves us with (to borrow the title of a book about him) an exquisite legacy of entomological art. Hudson produced all the illustrations that adorned his books, and many of these illustrations are now housed at Te Papa. According to his daughter Stella, Hudson's paintings started as pencil sketches of one side of the specimen, traced over to produce the whole drawing on card. This was painted over with watercolours applied with an extremely fine sable brush. In contrast to Helmore, these illustrations were made at publication size. For this book, we have reproduced Hudson's drawings in black and white with the kind permission of his family.

Ricardo Palma, entomology curator at Te Papa for forty years, is a world authority on lice. He used a compound microscope to view slide-mounted louse specimens but otherwise his illustration technique is broadly similar to Helmore's.

All the species found in this book occur in Aotearoa. The majority are endemic (i.e. are only found here), while some are native but not endemic (i.e. are found naturally in more than one country). The status of one species, the black field cricket, is uncertain. The insects are grouped in their orders, with the hemimetabolous orders preceding the holometabolous orders. (See the section on insect lifecycles for an explanation of these terms.)

Given that there are many thousands of species, we cannot cover everything in so small a volume, but we hope that we can give readers at least a hint of what makes the insects of Aotearoa so fascinating.

THE INSECTS

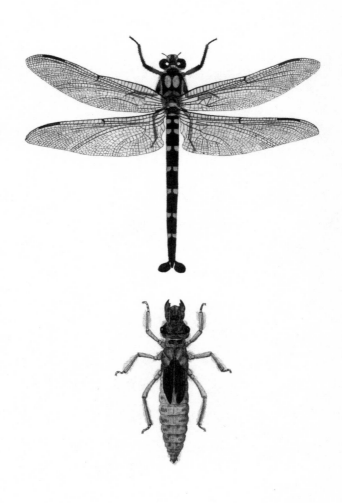

KAPOWAI
GIANT BUSH DRAGONFLY

Uropetala carovei

This is Aotearoa New Zealand's largest dragonfly species, although the closely related giant mountain dragonfly (*Uropetala chiltoni*) is only slightly smaller. They probably haven't changed much in appearance over millions of years and are loud and slow fliers compared with other dragonflies. As the name suggests, they prefer vegetation to open water bodies.

Description: Adult (above) body length is around 90mm and wingspan approximately 110mm across. Generally coloured black with pale yellow markings; possibly a form of disruptive camouflage when in flight. Naiads (below) look very different: no wings, colour brownish, and shorter and more heavily built.

Habitat and distribution: Kapowai prefer shaded forest or scrub habitat; naiads shelter in burrows in seeps and streams. Seen throughout much of Te Ika-a-Māui North Island and northern, western and southern Te Waipounamu South Island. Adults might be seen from November to May, most often in January and February.

Biology: Prey is taken in mid-flight. Huge compound eyes help these dragonflies lock on to a target even in a crowd of possible victims. A short burst of speed allows the dragonfly to hit its mark. Once the prey is secured, the dragonfly lands and consumes it in a leisurely fashion. Kapowai can take quite large insects, including bumblebees and butterflies, although smaller insects tend to make up the bulk of their diet. Adult males hold a territory near a suitable breeding site, patrolling it to defend against rivals, who may be grappled in mid-air. These dragonflies may also be seen resting on rocks and trees, basking in sunlight. Naiads are every bit as predatory as adults, emerging at night to hunt. This semi-aquatic larval stage may exceed five years.

Status in Aotearoa: Endemic

KIHITARA
REDCOAT DAMSELFLY

Xanthocnemis zealandica

Females of this species are noteworthy for having two colour forms. Naiads are territorial and wave their tails to deter rivals.

Description: Males (above) have bright red abdomens, giving rise to the common name redcoat damselfly. Some females (centre) also have a red abdomen, although the rear end is darker than in males. The second, more common, female colour form is dark on top with yellow sides. Females reach a maximum length of 39mm; males are slightly shorter. Naiads (below) are long and slender with prominent leaf-like gills on the rear end. Damselflies can be differentiated from dragonflies by the way they hold their wings when at rest: damselfly wings in line with the body, dragonfly wings out to the sides.

Habitat and distribution: Associated with pools, lakes, puddles and slow-moving streams, kihitara can tolerate a variety of water conditions from mountain tarns to swamps and brackish coastal pools. Seen from Te Tai Tokerau Northland down to Rakiura Stewart Island, they are typically on the wing between October and mid-March. Rēkohu Chatham Islands is home to a similar-looking related species, *Xanthocnemis tuanuii*.

Biology: Eggs are usually laid in submerged aquatic plant stems immediately after mating. Males may sometimes guard egg-laying females to deter rival males that might attempt to remove their sperm. The naiads are unusual among the Odonata (dragonflies and damselflies) for exhibiting territorial behaviour. They aggressively defend favoured perches on submerged stems or roots by twisting their bodies and waving their large tail gills; if this display is insufficient deterrence, they may attempt to strike each other with their gills. The naiad phase takes about two years to complete but may take three at higher altitudes. When the adult is ready to emerge, the naiad climbs out of the water on vegetation, the skin splits and the adult eventually pulls itself clear. Both adults and larvae are predators.

Status in Aotearoa: Endemic

23

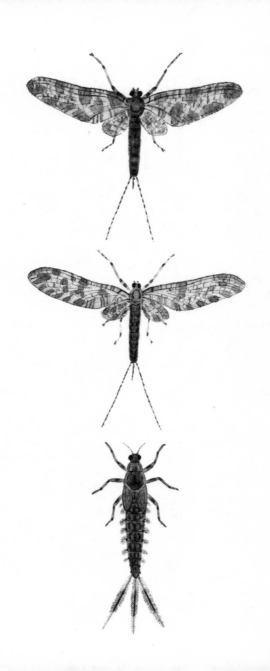

SMALL SWIMMING MAYFLY

Nesameletus ornatus

Mayflies spend almost all of their lives underwater among rocks on a streambed, usually a year. When conditions are right, they ascend to the surface to emerge and mate. In flight they keep their bodies vertical, with the three tails trailing behind, giving an overall impression of dancing in the air. The adult of this species has only two obvious tails and is particularly short-living in its wedding dance – only two days.

Description: Adults are 12–14mm long, with head and body dark brown with pale markings. Legs are yellowish, all with dark-brown bands, and wings have dark veins and yellowish patches marbled with black in males (above), and green in females (centre). The well-camouflaged nymphs (below) have an 11–18mm torpedo-like body shape with short antennae, large wing pads and seven pairs of small oval-shaped gills along the abdomen, all marked with a dark dot. Colour is initially almost transparent, changing to mottled grey and eventually almost black. They have three fringed tail filaments showing conspicuous dark bands, and short, banded legs.

Habitat and distribution: Found widely throughout Aotearoa New Zealand, especially in clear and smaller streams, up to an elevation of 1000m. Nymphs are most common in pools in stony or gravelly bush-covered streams in forests, on the edges among vegetation, where the current is slower.

Biology: Very active swimmers, the nymphs feed on biofilm (algae and other microorganisms) and plant detritus on the streambed. An abundance of this species suggests good habitat and water-quality conditions, especially if other mayfly or stonefly groups are also abundant. In December–January many moulted exoskeletons can be seen on dry stones close to the edge of the streambed.

Status in Aotearoa: Endemic

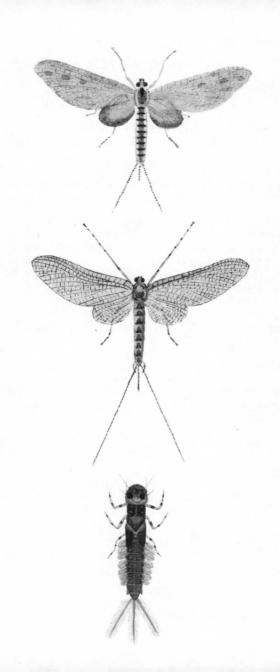

YELLOW DUN

Ameletopsis perscitus

Unlike most mayflies, this rather uncommon species is a predator of other aquatic invertebrates as a nymph or naiad. This is reflected in its large mouthparts, large (black) eyes and great agility. The sub-imago or dun (above) is called the yellow dun because of its bright yellow colour; they have radiant green eyes. It is believed that they mimic dead leaves when sitting still in this vulnerable stage.

Description: The imago (centre) is quite pretty with its 18mm-long pale body with dark markings and dots, with wings of a clear and shiny yellow. They have three tails: a short middle one and two longer outside. Nymphs (below) are well camouflaged: ocherous with dark markings and striped legs. The large round head carries conspicuous black eyes; the prominent mouthparts look like a sickle. The fringed tail filaments are rather short, as are the legs. The wing pads are unremarkable but the seven pairs of leaf-like gills along the body are large.

Habitat and distribution: Most abundant in bush-covered, gravelly streams with cool and well-aerated water under rocks throughout the year. Found throughout the country from Murihiku Southland hill streams to Ata Whenua Fiordland mountain streams and up Te Tai Poutini West Coast, and as far as the upper Te Ika-a-Māui North Island's rivers. The winged stage appears from the end of December to March. Adults are short-lived.

Biology: The carnivorous nymphs feed on smaller mayfly species and other small invertebrate larvae. They are good swimmers. Although not often found in high abundance, their presence is indicative of good habitat and water quality as they are very sensitive to water quality in streams. Larvae emerge onto dry stones over a rather long period of four months between December and March.

Status in Aotearoa: Endemic

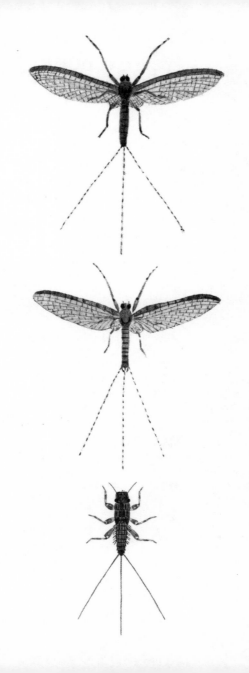

DOUBLE GILL MAYFLY

Zephlebia dentata

The larvae of this species are very common; a good food source for fish. They crawl out of the water for a while before the adults emerge. The flight pattern of the adults shows a typical up-and-down movement.

Description: Nymphs (below) have a flattened body form, which is helpful when holding on to a surface in water currents. They have sharp spines on some abdominal segments, but what gives them their common name is the double row of abdominal gills on both sides. The wings of the adults (female imago above, female subimago centre) are uniformly brown with blackish shades.

Habitat and distribution: Larvae are very common in many hard-bottom or soft-bottom, bush-covered and farmland streams, especially in Te Ika-a-Māui North Island, and in Te Waipounamu South Island's mountains and bushy areas. They prefer cool, well-aerated water and can occur in high numbers under a single rock. This species is abundant throughout the year in all development stages.

Biology: Larvae are rarely seen swimming or walking. Like other leptophlebiid mayflies, they feed by scraping diatom algae and other organic matter from stone surfaces. While they are less sensitive to water quality than other species, they are still a good indicator of a healthy stream. The mouthparts of the adults are dysfunctional and the digestive system is filled with air, as at this stage they do not feed and live for only a short time in order to reproduce.

Status in Aotearoa: Endemic

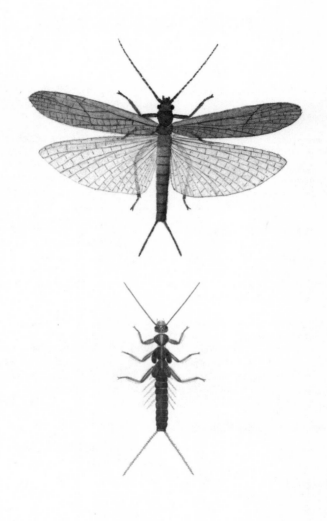

LARGE GREEN STONEFLY

Stenoperla prasina

The large green stonefly is quite iconic with its bright colouring. Despite the name, they sometimes appear yellow. The larvae are one of the most sensitive aquatics to stream quality; if you find one, you can be assured that it came from a healthy ecosystem. Some related species from higher altitudes are wingless.

Description: Adults (above) can be up to 20–30mm long, of a bright green or sometimes yellow colour. At rest the wings are neatly folded along the back, resembling a roll of leaves. The eyes appear red. Both adults and larvae (below) have a typical elbow-bend in the legs. In larvae the future wings are hinted at by wing buds on the back. The abdomen is long and segmented, typically brownish with two tail filaments. Similar to mayflies (which usually have three tail filaments), larvae have gills along both sides of their body, which can appear pink.

Habitat and distribution: Found throughout Aotearoa New Zealand in unmodified, bush-covered, cold stony-bottomed streams from sea level to the alpine zone.

Biology: Females scatter their eggs in the water, where they become attached to the stream bottom, surrounded by a protective jelly. The small hatching nymphs feed on a range of streambed detritus. Larger nymphs, however, are known to be predators of other stream invertebrates. The adults, being weak fliers, often rest during the day in dense vegetation along the streambanks. They are attracted to light, though, so they can sometimes be found closer to dwellings.

Status in Aotearoa: Endemic

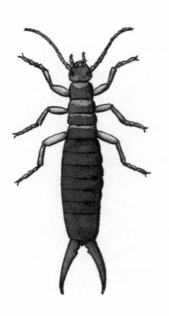

NGĀ MATĀ
SEASHORE EARWIG

Anisolabis littorea

Despite its name, an earwig normally doesn't crawl into human ears, although they do like to hide in narrow, dark and humid places. This native earwig will happily crawl under seaweed. It is much larger than the European species and, unlike them, is flightless. Like other earwigs, ngā matā care for their young during development, but would also eat them when times are tough.

Description: About 35mm long, although this varies because their abdominal segments can be stretched and contracted like a telescope. The body is blackish-brown with brown-yellow legs. It has two light-brown spots on its head, close to the inside of each eye. The abdomen is wider in the middle and generally wider than the head, with the pincers characteristic of earwigs at the rear end being asymmetrical in males.

Habitat and distribution: Native to eastern Australia and Aotearoa New Zealand, ngā matā are found in moist and confined places on beaches, near the seashore, under stones and driftwood throughout Te Ika-a-Māui North Island and Te Waipounamu South Island as far south as Ōtepoti Dunedin.

Biology: Larvae go through five instars before becoming adults. At all stages they are active at night and hide from light and in the daytime (called negative phototaxis), usually in damp, sheltered places. They are predators of millipedes, flies and isopods, but their main prey are sand hoppers. The scary-looking pincers at the end of their tails, which are only used in defence in other species, are here used to hunt and cut their victims. The females protect and clean their offspring.

Status in Aotearoa: Native

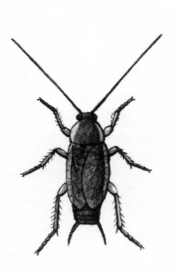

NATIVE BUSH COCKROACHES

Family Blattellidae and others

Cockroaches are probably the most misunderstood insects. Aotearoa New Zealand's native bush cockroaches live outside, tearing apart organic matter and helping to make humus, using bacteria in their gut to break down wood fibres. They are no household pest!

Description: The nymph that emerges from the egg looks very similar to the adult, just smaller. The largest native species, *Maoriblatta novaeseelandiae* can reach 30mm. Like the nymphs, adults are flat and many species are wingless, which is helpful for squeezing through leaf litter, under loose bark and in other tight spaces in rotting plant material. Native cockroaches range from light brown to black. Antennae are as long as the body. Two small appendages on the abdomen, known as cerci, act as sensors which allow an extremely fast escape from predators. The mouth of a cockroach can move from side to side.

Habitat and distribution: More than thirty-one native cockroach species are found throughout Aotearoa New Zealand, with close relatives living in Australia and New Caledonia. They live outside in logs and leaf litter, and under loose bark. They love damp, dark habitats, and only come inside by accident.

Biology: Cockroaches are survivors, able to live for long periods without food and water. *Celatoblatta quinquemaculata* can even freeze and become active again once the temperature increases, while *M. novaeseelandiae* secretes chemicals that actively repel predators. A mature female cockroach will lay hundreds of eggs in her lifetime; generally these are compiled in egg cases, but some species give birth to live nymphs (viviparous). Before becoming fully reproductive adults, nymphs moult (shed their body casing) several times. This process can take between a few weeks to a year to complete.

Status in Aotearoa: Endemic

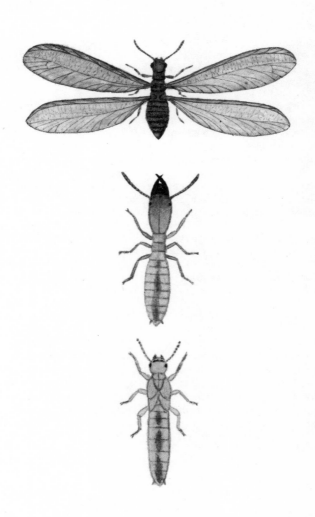

WETWOOD TERMITE

Stolotermes ruficeps

Termites live in colonies with a queen. A colony usually consists of between 50 and 250 termites, most of which are nymphs, but if the host material is suitable a colony may grow to about 3000 individuals and sub-colonies connected to the initial one may be formed. The nymphs construct surface tunnels of cemented soil, faecal pellets, wood particles and other detritus to protect themselves when passing from an established colony to new feeding grounds. Nymphs become either workers or soldiers.

Description: Adults (above) are dark brown to black and about 12mm long. The head is almost spherical with long, slender antennae and a narrower segment immediately behind the head. The dark wings are about twice as long as the body and are folded flat over the abdomen when at rest. Nymphs are white to pale yellow with external wing buds. Soldiers (centre) do not have wing buds; their long heads are dark at the front but orange at the rear, with jaws about one-third as long as the head.

Habitat and distribution: Colonies are found in forests throughout Aotearoa New Zealand, in dead and rotting trees, branches, logs and stumps. They prefer native kauri, rimu, northern rātā, beech and *Podocarpus* as well as exotic macrocarpa, eucalypts, willow and pine.

Biology: Winged adults swarm in autumn, and rarely fly further than 30m. Most shed their wings immediately after the first flight and move to shelter in cracks and holes. The first eggs are laid in the following spring and take 33–35 days to hatch. The first nymphs to emerge are fed by the parents and later function as a worker caste (below); some develop into soldiers. Growth of a colony may continue for three or more years before mature nymphs become winged, reproductive adults and swarm.

Status in Aotearoa: Endemic

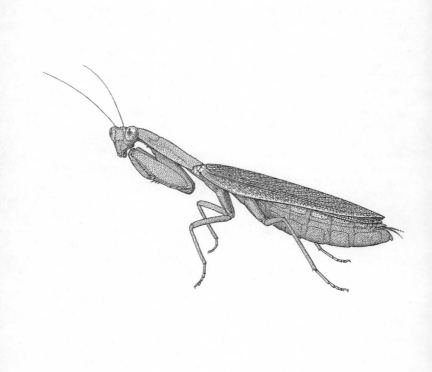

NEW ZEALAND MANTIS

Orthodera novaezealandiae

The New Zealand praying mantis is probably our most efficient insect predator. Its weapon – the forelegs – is always carefully cleaned and folded when not used, giving rise to the name. Once a mantis has drawn a bead on a prey, its forelegs grab it with great speed and precision; the victim barely stands a chance.

Description: About 40mm long, winged and usually a bright apple-green colour, although some variation in colour does occur. The forelegs are long and spiny. Males are slender; females have a much bigger abdomen, especially when pregnant. Both sexes have large bulging eyes, which give the face a triangular shape. A characteristic bluish spot on the inside of the front legs makes it very easy to distinguish them from the only other mantis species in Aotearoa, the introduced South African springbok mantis.

Habitat and distribution: Can be found in lowland throughout Te Ika-a-Māui North Island and Te Waipounamu South Island.

Biology: Mantises often move the head from side to side when tracking prey. They can run fast and are able to fly. The praying mantis is renowned for sexual cannibalism, whereby the females eat males during mating (or even before they get a chance to try). However, the New Zealand mantis is rarely cannibalistic, preferring to snack on flies. Eggs are laid in two neat rows, covered in a foamy egg case called an ootheca. The foamy substance hardens over time, protecting the eggs inside. Oothecae are laid on the surface of branches and tree trunks, but also on the sides of fences and houses.

Status in Aotearoa: Endemic

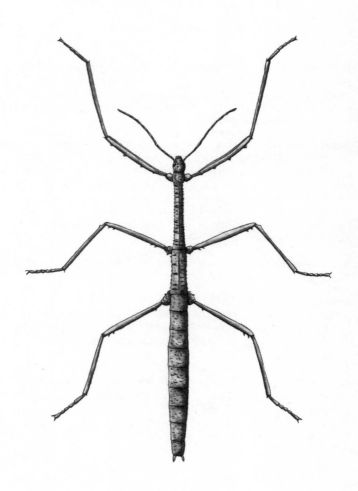

BRISTLY STICK INSECT

Argosarchus horridus

This is the largest stick insect in Aotearoa New Zealand and our longest insect overall. As is the case with a number of our stick insect species, males are optional, with some populations consisting entirely of females.

Description: Females are typically around 12–15cm long, but can approach 20cm. Males are around 10cm in length and more slender than females. Colouring is mottled grey or light brown in females; males are shaded in browns and greens. The thorax in both sexes is bristly, with males having fewer but longer spines than females. The name *horridus* refers to these spiny bristles.

Habitat and distribution: Can be found throughout Te Ika-a-Māui North Island, but less common in Te Tai Tokerau Northland. On Te Waipounamu South Island they are more likely to be seen in coastal and lowland areas. Also found in Rēkohu Chatham Islands. May sometimes appear in gardens, feeding on roses.

Biology: After hatching from an egg that resembles a plant seed, *Argosarchus horridus* goes through six nymph stages before maturing to an adult. Some populations may contain both sexes, while others, like those found on Rēkohu, may consist entirely of females. In this situation, females reproduce by parthenogenesis, laying viable eggs that are unfertilised by males. Their offspring is always female. Like other stick insects, *Argosarchus horridus* is a master of disguise, looking very much like a lichen-covered twig. They also minimise their exposure to predators by being nocturnal. Bristly stick insects feed on a variety of plants, including *Plagianthus regius*, *Lophomyrtus bullata* and *Muehlenbeckia australis* as well as introduced species such as raspberries and roses.

Status in Aotearoa: Endemic

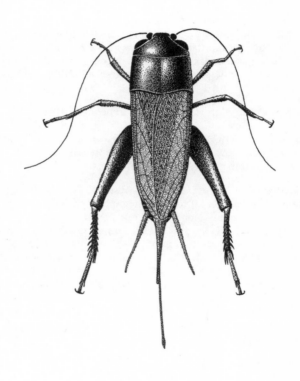

PIHAREINGA
BLACK FIELD CRICKET

Teleogryllus commodus

Pihareinga are more often heard than seen and are regarded as a major pest. In suitable conditions local populations can reach very high levels and cause significant damage to pasture.

Description: These insects are black, and adults are 25–35mm long. Their powerful hind legs are a hallmark of the Orthoptera order to which they belong. Adults have wings, held flat over the abdomen when not in use. Nymphs are similar in appearance to adults but the wings are underdeveloped or absent.

Habitat and distribution: Found throughout most of Te Ika-a-Māui North Island and upper Te Waipounamu South Island; has also been found in Waitaha Canterbury. Mostly found in grassland habitat between spring and autumn. Pihareinga use cracks in dry soil as daytime refuges.

Biology: Eggs are usually laid between February and May, hatching in spring. Nymphs mature rapidly. Flight is used primarily for dispersal to new areas in February and March, with jumping the preferred means of escaping danger. The chirping calls of males to females are often heard on warm, dry summer nights. Males make their calls by rubbing a hardened area on the leading edge of one forewing against teeth on the underside of the other as they open and close their wings. Males also fight. Two males will initially assess one another through antennal contact and a display of mandible flaring, progressing to wrestling and biting if one of the males doesn't withdraw. Pihareinga are pasture pests and are particularly problematic in hot, dry summers when dry, cracked ground provides plenty of daytime shelter.

Status in Aotearoa: Uncertain; considered to be introduced from Australia, but whether it was self-introduced or not is unclear

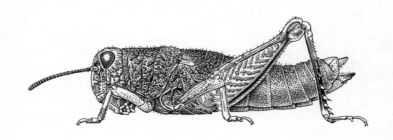

ROBUST GRASSHOPPER

Brachaspis robustus

Like most grasshoppers, this species can jump, but it is terrible at landing. A declining population and a very small range have led to this species being one of only two grasshopper species protected under the Wildlife Act 1953. It is also one of the first insects in the world to have a purpose-built reserve, of 440 hectares near Tekapō and surrounded by a predator-proof fence, created for its protection.

Description: Three colour forms are known for these somewhat sturdily built grasshoppers. Most specimens are grey; others are orange-brown or, very rarely, black. Females can be up to 42mm long, while males reach barely half that length. Their minute wings are functionally useless.

Habitat and distribution: Known only from three river catchments in Te Waipounamu South Island's Mackenzie Basin, where it lives in rocky areas in preference to grass.

Biology: The lifespan is around two years. Females lay about thirty eggs. Nymphs typically emerge in summer, overwintering in sometimes sub-zero conditions before reaching maturity the following summer. They are general herbivores, eating mosses, lichens and leafier plants. Like most New Zealand grasshoppers, they are flightless. According to grasshopper expert Tara Murray, male robust grasshoppers can leap as far as 1.5m, with the larger females capable of shorter distances. Landing, however, is clumsy, with the insect often hitting the ground on its back or belly with an audible 'thock'. Instead of leaping to safety as a first defence, these grasshoppers rely on standing absolutely still while their colouring helps them blend into the background. Despite being difficult to see, they remain vulnerable to predators such as hedgehogs and rodents that find their prey through scent.

Status in Aotearoa: Endemic

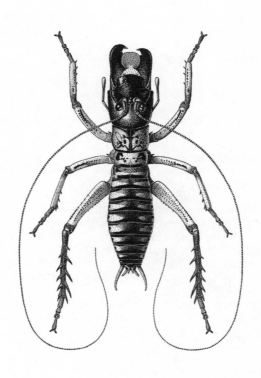

WELLINGTON TREE WĒTĀ

Hemideina crassidens

When it comes to sex, size – specifically male head size in this case – doesn't always matter. Some males possess large, imposing heads with impressive mandibles, and so are better equipped to guard entrances to tree cavities (called galleries), where they maintain harems of females. Smaller-headed males use other strategies to mate.

Description: Fully grown large-headed males may reach 70mm long. Females are easily identified by having a slightly curved, sword-like ovipositor on the rear of the abdomen. Colouring is similar in both sexes. The head is red-brown with long antennae, while the first part of the thorax is covered with brown to black saddle-like pronotum. The abdominal segments have alternating bands of dark brown or black and yellow or light brown. The hind legs are armed with strong spines on the tibiae.

Habitat and distribution: Found in tree cavities in the lower Te Ika-a-Māui North Island and the north-west of Te Waipounamu South Island. They may sometimes make use of artificial objects that provide similar living conditions.

Biology: These insects live in social aggregations in galleries, which may originally be abandoned holes made by other insects such as the pūriri moth (*Aenetus virescens*). Males, particularly large-headed individuals, guard harems of females, although juveniles, including males, may also be present. Smaller-headed males may guard harems of their own when the gallery entrance is too small to permit bigger males to enter. Smaller males may also mate with females foraging in the open. Although herbivorous, tree wētā are known to scavenge dead insect carcasses. They use stridulation to create sound, rubbing pegs on the hind femur against ridges on the body. Males may call to attract females, while both sexes can make defence calls when threatened or an eviction call when a wētā is being evicted from a gallery.

Status in Aotearoa: Endemic

COOK STRAIT GIANT WĒTĀ

Deinacrida rugosa

Although this insect is large and heavily armoured, that does not save it from introduced mammalian predators such as rats and stoats. However, thanks to several successful projects translocating giant wētā to rat-free environments, this species is making a comeback. Like all giant wētā (members of the genus *Deinacrida*), this species is absolutely protected under the Wildlife Act 1953. It is illegal to catch, injure or hold specimens (even parts of specimens!) without legal authorisation.

Description: Tan to medium brown in colour, the bulky-looking body is covered with heavy-looking plating. Females are larger, reaching about 70mm long, and have a prominent ovipositor on the rear end. Males are up to 50mm long.

Habitat and distribution: Open grassland or grassland/shrubland on rat-free islands in Te Moana-a-Raukawa Cook Strait; has also been reintroduced to Mana Island, Matiu Somes Island and the Zealandia eco-sanctuary in Te Whanganui-a-Tara Wellington. The population in Zealandia, sourced from Matiu (also the home of a translocated population), is the first on the mainland in over a century. Translocation projects like this show that this species can survive well in suitable predator-free environments.

Biology: Nocturnal, by day they hide in temporary refuges excavated at the bases of vegetation or under logs or stones, emerging at dusk to feed on foliage. The lifespan is around two years, with females laying around 200 eggs in their second year. Despite the heavily armoured appearance, they are very vulnerable to predation by rodents, particularly rats. The very strong scent of females allows males to locate them, but also makes it very easy for predators to find them.

Status in Aotearoa: Endemic

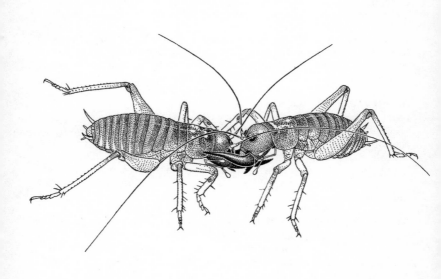

TUSKED WĒTĀ

Motuweta isolata

Males of this species are noted for their very obvious mandibles, giving rise to the name tusked wētā. The territorial males use their mandibles in shoving contests, each attempting to flip the other over.

Description: This species can reach 90mm in length. Males have large, brown mandibles that project forward. Females lack the tusks but have an ovipositor on the rear end. Colouring is often red-brown, particularly around the head region. Legs are tan.

Habitat and distribution: Found only on the Mercury Islands, near Coromandel, and Repanga Cuvier Island to their north, they live in burrows in the ground. Originally discovered on a single island (Atiu Middle Island) in the Mercury group, it appears to have gone extinct there, with no sightings since 2001. Fortunately, the decline in population had been noticed in the 1990s and a captive breeding programme was instituted. This was sufficiently successful to allow the establishment of new populations on two other islands in the Mercury group. There have been further introductions to other islands in the group, as well as to Repanga to the north. The captive breeding programme has undoubtedly saved this species from immediate extinction, but the entire population of tusked wētā is descended from a single male and two females, so genetic diversity is very limited. *Motuweta isolata* is a fully protected species under the Wildlife Act 1953.

Biology: Eggs are laid in small depressions in the soil. Captive rearing indicates that these insects take about 18 months to mature. They spend a lot of time in their burrows, only emerging to feed on moonless nights, possibly as a way to avoid predation by tuatara and other reptiles. Unlike most wētā, this species is largely carnivorous, feeding on worms and insects.

Status in Aotearoa: Endemic

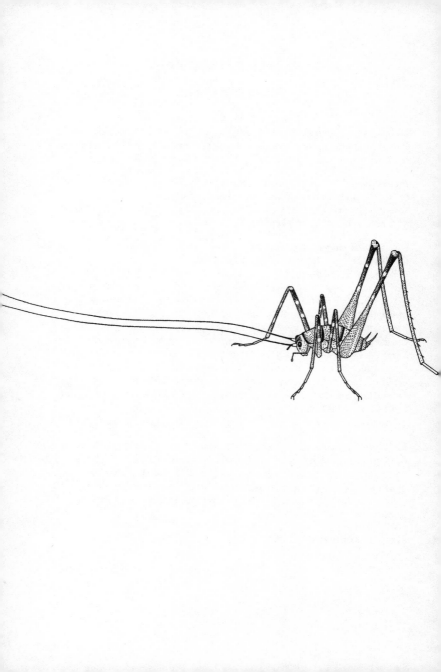

CAVE WĒTĀ

Pachyrhamma spp. and others

Some cave wētā do indeed live in caves, but as a group they make use of a far wider range of habitat types including tree hollows, leaf litter, under rocks and logs or even in basements. As members of the family Rhaphidophoridae, they are not particularly closely related to true wētā, which all belong to the family Anostostomatidae.

Description: Most species of cave wētā have extremely long antennae and rather spindly legs – an easy way to tell them apart from true wētā. Size varies from 5mm in length to species of *Pachyrhamma* – an example is illustrated here – with a 50mm body length carrying antennae 250mm long and hind legs measuring 110mm.

Habitat and distribution: Found throughout Aotearoa New Zealand. Entomologist Mike Meads notes that 'any dark, damp and cool hole or space that has ready access to the outside is frequently used'.

Biology: Unlike true wētā, cave wētā don't produce sound, and lack special hearing organs called tympana. However, they can pick up ground vibrations through their feet, and some airborne vibrations via cerci (small projections on the rear of the insect) and their incredibly long antennae. They are excellent jumpers, and leaping to safety is their primary defence mechanism. This tactic doesn't always work, however. A Nelson cave spider (*Spelungula cavernicola*) can grab a cave wētā and then drop on a dragline of silk, robbing the wētā of a surface to launch from. Cave wētā sometimes congregate in large numbers by day before emerging to forage singly at night. They have minuscule mouthparts, and rather than feeding on foliage they consume lichen, fungi and moss. They will also scavenge dead insects but are not known to actively predate live ones.

Status in Aotearoa: Endemic

KIKIHIA
CICADA

Kikihia spp.

Aotearoa New Zealand is home to sixteen described species of the genus *Kikihia*, but there could be as many as twenty-eight species present. They are believed to be descended from New Caledonian colonists and to have diversified rapidly around 3–5 million years ago, possibly in association with the uplift of Kā Tiritiri o te Moana Southern Alps.

Description: Adults (above) of most *Kikihia* species are at least partly green in colour, and are typically smaller and more slender than other Aotearoa cicada genera such as *Amphipsalta* and *Maoricicada*. Nymphs are wingless, pale or brown and have forelimbs modified for digging.

Habitat and distribution: Found right across Aotearoa, including more distant areas such as Rangitāhua Kermadec Islands. The only species found outside Aotearoa lives on Norfolk Island. Habitats vary by species. For example, *K. scutellaris* prefers lowland or low montane forest; *K. rosea* is most often seen on grasses and shrubs from lowlands to the subalpine zone. Adults are usually seen in late summer, but in some species may be present in low numbers during spring and autumn.

Biology: Lifecycle length for *K. ochrina* is estimated at around three years; it is unknown for most species. Eggs are laid in branches, often in summer, with nymphs emerging the following spring. Nymphs burrow into the ground, where they use their straw-like mouthparts to feed on sap from plant roots. When ready to emerge, the final-stage nymph climbs a suitable vertical surface nearby (often a tree) during darkness. Like other cicadas, only males sing, accomplished by contracting and releasing a membrane called a tymbal (shown in the lower illustration), located at the front underside of the abdomen. The sound is amplified by air sacs. Each species of cicada has its own song.

Status in Aotearoa: Endemic except for one species that is endemic to Norfolk Island

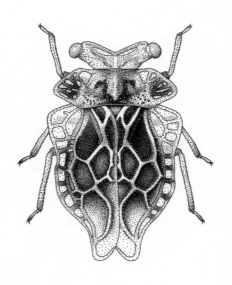

MOSS BUGS

Family Peloridiidae

These tiny, ornate-looking bugs are sometimes referred to as living fossils, belonging as they do to the suborder Coleorrhyncha, a group which has a fossil record dating back to the Permian period (nearly 300 million years ago). Their current-day distribution across the southern forests of the world suggests that their evolutionary history is tied to the break-up of the ancient supercontinent of Gondwana. Aotearoa New Zealand is home to thirteen species of moss bug, representing about 36 percent of the world's diversity for this family.

Description: Small (2–4mm long) and typically flat. The upper surface is usually highly ornamented. The eyes are on small projections sticking out from each side of the head. They lack hindwings and cannot fly.

Habitat and distribution: Found on Te Ika-a-Māui North Island, the west of Te Waipounamu South Island and on Rakiura Stewart Island, in wet moss and usually in association with trees. Adults typically emerge between December and March.

Biology: Moss bugs go through five nymph stages before reaching maturity. As adults and nymphs are rarely seen during winter, it is thought they may overwinter in the egg stage. Little is known of their behaviour. Given their inability to fly because of their lack of hindwings, they are likely to have poor dispersal ability. When conditions get drier, it is assumed they move deeper into the moss in search of greater levels of humidity.

Status in Aotearoa: Endemic

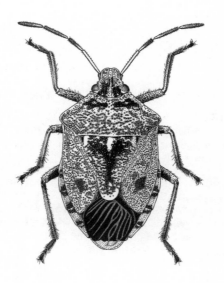

BROWN SOLDIER BUG

Cermatulus nasalis nasalis

These insects may have a similar shape to familiar garden pests such as the green vegetable bug, but they are predators – instead of feeding on plants, they stab their straw-like mouthparts into insect prey and suck the juices out.

Description: Shield-shaped bodies, with adult females reaching 12mm in length. Males are about half the size of females. Adults are shades of brown with black markings. Nymph stages look black from above with yellow or white markings about halfway along each side.

Habitat and distribution: Brown soldier bugs are found on both main islands and may be observed hunting for prey on a wide variety of plant species; they are also known from Australia and Timor. Two closely related endemic subspecies, alpine (*C. n. hudsoni*) and Turbott's (*C. n. turbotti*) brown soldier bugs, are found in Te Waipounamu South Island alpine areas and from kānuka (*Kunzea ericoides*) on Manawatāwhi Three Kings Islands, respectively.

Biology: Brown soldier bugs have a lifecycle consisting of an egg stage and five nymph stages before reaching adulthood. All three subspecies belong to the family Pentatomidae, more commonly known as shield bugs because of their shape, but are also called stink bugs as they can emit strong-smelling chemicals to deter predators. The brown soldier bug feeds on caterpillars of several important pests as well as slugs and eucalyptus tortoise beetle (*Paropsis charybdis*) larvae. Unfortunately for admirers of monarch (*Danaus plexippus*) and red admiral (*Vanessa gonerilla*) butterflies, their caterpillars are also vulnerable to predation by this species.

Status in Aotearoa: Native to Australia, Aotearoa and Timor

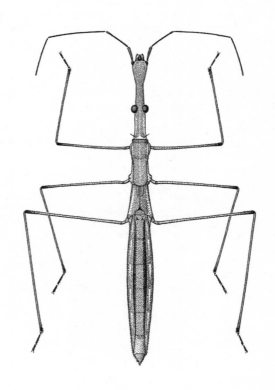

WATER MEASURER

Hydrometra strigosa

The genus name *Hydrometra* translates as water measurer, a reference to their slow, deliberate stepping across water surfaces.

Description: Brown or yellow-brown in colour. Legs are long and fine. Their slender bodies are around 10–15mm long. The head is also long, with the eyes set back a little over halfway.

Habitat and distribution: They prefer quiet bodies of water, usually fresh, although they may sometimes be found on salt or brackish water. Mostly seen in the upper half of Te Ika-a-Māui North Island, but also recorded from the Whakatū Nelson and Waitaha Canterbury regions.

Biology: With their slender form and unhurried movements, these insects superficially resemble stick insects (order Phasmotodea). However, aside from the obvious size difference, they are very different animals. Stick insects are masters of blending in and are herbivores that chew their food; water measurers do not rely on camouflage and are predators with piercing and sucking mouthparts. Little is known about the biology of *Hydrometra strigosa*, but it is likely to be similar to that of its relatives elsewhere. With their slender legs and bodies, these insects can walk on water without breaking surface tension. Their delicate appearance belies the fact that they are grimly effective predators. They lurk on the water's surface or on algae protruding through it, waiting for insects to fall on the water's surface or approach the surface from below. Vibrations caused by prey movement are picked up through the water measurer's legs. It then spears the prey with the tip of its snout, holding it fast while inserting a pair of stylets into the body to tear up the flesh before consuming the fluids.

Status in Aotearoa: Native; first described from Australia and is recorded from Tahiti, New Caledonia, Norfolk Island and Vanuatu

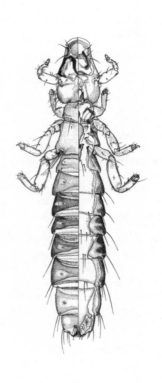

FEATHER LICE

Family Philopteridae

These tiny parasitic animals are completely dependent on their bird hosts for survival. They are also model animals for coevolutionary studies: both bird and louse often share an evolutionary association, with both co-speciating together.

Description: Can vary greatly in appearance, but are all wingless and flat with legs adapted to hold on to feathers. Generally small (under 5mm long), but *Harrisoniella hopkinsi* can approach 10mm in length. Members of the genus *Naubates* are elongate and dark brown. The left side of the illustration shows the feather lice from above and the right shows the underside.

Habitat and distribution: Geographic ranges for lice are essentially the same as their hosts. *Naubates thieli* (shown here) is only found on the providence petrel (*Pterodroma solandri*), a vagrant visitor to Aotearoa New Zealand.

Biology: As the name suggests, feather lice feed on feathers. The egg is followed by three nymph stages before adulthood. Feather lice are obligate parasites, meaning they can only survive on their bird hosts, depending on them for warmth, shelter and food. Generations of lice may live their entire lives on the same bird. If the host dies, so do the lice, unless they can quickly transfer to a new host; this requires close physical contact between suitable hosts. Just as lice have become adapted to life as a parasite, so too have hosts developed adaptations to control their numbers. This evolutionary 'arms race' has led to the coevolution of both birds and their lice, to the point where most species of louse live exclusively on one or very few host species. Consequently, the phylogenetic histories of lice can often help elucidate those of their hosts. If the host species goes extinct, so do any louse species that live on it. For example, the aptly named *Rallicola extinctus* is known only from museum specimens of the extinct huia (*Heteralocha acutirostris*).

Status in Aotearoa: Depends on whether the host is endemic, native or introduced

THRIPS

Cleistothrips idolothripoides

This species is the only member of the genus *Cleistothrips*. Little is known of its biology, but it is thought to feed on fungal spores. Thrips is both the plural and singular name for this group of insects.

Description: At over 4mm in length, these are comparatively large thrips. The body is dark brown and cigar shaped. The near half of each leg is dark, while the furthest half is yellowish. As with other thrips, the wings have a superficial resemblance to feathers.

Habitat and distribution: Found throughout most of Te Ika-a-Māui North Island and upper Te Waipounamu South Island. It has been found on both native and exotic vegetation. Specimens have been collected in all four seasons, suggesting that adults may potentially be present year-round.

Biology: Thrips are characterised by their asymmetric mouthparts and feathery-looking wings. Little is known about the biology of *Cleistothrips idolothripoides*, but the structure of its mouthparts suggests it may feed on fungal spores. It is assumed to have a similar life-history to other members of the family Phlaeothripidae. This includes two actively feeding larval instars followed by three nearly inactive and non-feeding pupal instars. As a group, fungus feeding is most common, but flower feeding, gall forming, predation and even parasitism are known. Some species of thrips are pests of crops, and a few are known to be vectors for plant viruses. However, some species are useful pollinators. Females are capable of laying eggs whether they have mated with a male or not. In this species, adults are thought to breed in hollow twigs.

Status in Aotearoa: Endemic

STRIATED ANT

Huberia striata

Most species of ant seen in Aotearoa New Zealand belong to genera found in other countries. The genus *Huberia* is the exception, being only known from Aotearoa, which makes it our only endemic genus of ants.

Description: Workers are usually over 4.5mm long, with larger specimens reaching 5mm; however, workers from newly established nests may only reach 4mm. The first segment of the abdomen is armed with a pair of sharp spines. Reproductive females are much larger than workers and can be about double worker size. Males are always blackish in colour, while worker and female coloration varies widely. Workers range from bright reddish yellow to nearly jet black; the female colour pattern is darker than that of workers.

Habitat and distribution: Very widely distributed in Aotearoa, these ants nest in soil, often under stones in cooler places, although they may also nest in rotting wood in northern parts of the country.

Biology: These ants are not aggressive. Groups from different localities can be combined into a single nest without conflict, robbery or slavery as other ant species may do. Nests are very large without definite colony limits. Striated ants have mutualistic relationships with several other invertebrates. Scale insects, young planthoppers or mealybugs that are sometimes found in ant nests are protected by the ants, which benefit from feeding on the waxy secretions they produce when feeding on plant roots. Native woodlice can also be seen in some nests, possibly helping to keep the nest clean by feeding on ant droppings. They are clearly of value to the ants, which will carry them to safety if the colony is disturbed. Striated ants are generalist omnivores, foraging cooperatively in subterranean or leaf litter.

Status in Aotearoa: Endemic

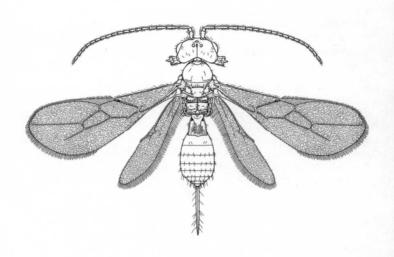

BRACONID WASP

Aphaereta aotea

This species is a member of the family Braconidae, second only to Ichneumonidae in terms of species diversity within the order Hymenoptera. Their larvae develop inside the maggots of several species of fly.

Description: These wasps are small (up to 4.5mm long, including a long ovipositor in females) with a glossy dark red-brown to black body. The wings are glassy with a brown tinge. Legs are yellowish near the body, darkening further away.

Habitat and distribution: Found throughout most of Aotearoa New Zealand.

Biology: Female braconid wasps lay eggs on or in the larvae of other insects. After hatching, the larvae feed on their hosts, often weakening or killing them. Braconids typically have a narrow range of suitable host species, making some excellent biological control agents for specific pests. Despite being endemic to Aotearoa, this species was first described in Australia, where live specimens had been imported as a possible biological control agent for dung-breeding flies. In Australian testing, males emerged first and mated with females almost as soon as they appeared. Females were ready to lay eggs within half an hour of emerging. After homing in on host larvae, females then briefly paralysed their targets by stinging them before commencing egg-laying. The number of eggs laid on a single host may depend on the size of both the host and the wasp. Larger wasps may have more eggs to lay, and larger hosts may receive more eggs. Wasps emerge from the pupal stage of the host, with larger hosts more likely to survive the infestation. The time taken for the wasp to complete its lifecycle from egg to adult emergence is hastened by warmer temperatures.

Status in Aotearoa: Endemic, but also introduced to Australia

SPIDER-HUNTING WASP

Sphictostethus fugax

This species is a member of the parasitoid family Pompilidae, which are famous for their ability to overcome spiders that are much larger than they are. As they hunt, they move with quick, darting motions and twitching antennae. Once a suitable victim is found and subdued, the wasp drags it back to its nest.

Description: Up to 15mm long, this wasp has a glossy, red-brown appearance with an eye-catching golden area around the middle of the body. Wings are yellowish. The larvae can reach 17mm in length; they look like a segmented grub with a broad rear half.

Habitat and distribution: Found throughout Aotearoa New Zealand, often in close proximity to suitable nesting sites, which are typically abandoned beetle pupal chambers. Wasps can be seen from spring to autumn.

Biology: These wasps are solitary rather than social and only females hunt spiders. They search for prey in leaf litter, logs and webs on tree bases. Spiders are chased into the open and stung several times, leaving the spider completely paralysed. The wasp then searches for an empty beetle pupal chamber (usually in a tree), before returning to the spider and hauling it there. Moving a spider over rough ground can be difficult, but entomologist GV Hudson noted that spider-hunting wasps that encounter rugged terrain sometimes temporarily abandon their captured spider to work out an easier path. Once the spider is in the nest, it is sealed inside with an egg. The hatching larva feeds on the still-living but paralysed spider until it is time to pupate. Eventually, a new wasp will emerge. Only the larval stage is carnivorous; the adults feed on nectar.

Status in Aotearoa: Endemic

PUENE
DOBSONFLY, TOEBITER

Archichauliodes diversus

This dobsonfly species is the only representative of its family in Aotearoa New Zealand, where it is one of the largest and most spectacular aquatic insects. Larvae have gills that look like extra pairs of legs or tentacles, along with large pincer-like mandibles that gave them the misleading name toebiter.

Description: The cylindrical eggs change their colour from light yellow to dark brown over about a month. Larvae (below) are around 2.2mm when they hatch and can grow to around 38.5mm. Resembling a centipede that wriggles through the water, they have a flat, elongated head and first segment of the thorax, both of which are hard and dark, almost black compared with the softer pale abdomen that carries eight pairs of gills. Adults (male above, female centre) have large, round eyes, long antennae and very large, strong, slightly curved speckled wings that are normally folded behind the back. Pupae are similar to adults but with a bigger build and undeveloped wings.

Habitat and distribution: Larvae are common in hard-bottom streams with sheltering bush vegetation and moderate to good water quality. The third larval stage (prepupa) migrates towards the water edge or banks, from early July to late January, when water levels are at maximum height.

Biology: Puene are predators of other aquatic insects, such as mayflies and caddisflies, which they ambush and catch using their large mandibles. The prepupa needs the soils to be saturated for the next lifecycle stage (pupa) to occur. The pupal stage lasts around 20–24 days from late October to February, with males taking longer than females. Adults are nocturnal, with a short lifespan of 6–10 days. Females lay hundreds of eggs on the banks and rocks of streams.

Status in Aotearoa: Endemic

NEW ZEALAND ANTLION

Weeleus acutus

The larvae of this species use a deadly combination of slippery slopes, well-aimed soil particles and vicious, venomous jaws to ensure that they catch their prey.

Description: Adults are slender, largely grey-black insects reaching about 40mm in length; their net-veined wings are longer than the body. Larvae are almost pear-shaped with prominent jaws; most of the body is hidden from view at the bottom of their prey-trapping pits.

Habitat and distribution: Found across Aotearoa New Zealand, adults hide in foliage by day and emerge to hunt after dark. They are usually seen in mid to late summer. Larvae make distinct conical pits in loose, dry soil or sand in sites sheltered from rain but exposed to sun.

Biology: Eggs are laid in loose soil or sand. The larva creates a conical pit with steeply sloping sides, using its abdomen to dig a circle marking the outer edge of the pit, then flicking earth clear using a flat area on its head. The pit is expanded as the larva grows. It sits in wait at the bottom with only its jaws and head visible. Frequently preying on ants (hence the common name), it will take almost any insect or spider that comes its way. Should an unwary insect enter the pit, the combination of loose footing and a steep slope makes escape difficult. The antlion also makes things harder by flicking earth at its prey to try to knock it down. Once it is close enough to bite, the prey's struggles are soon over – the antlion's powerful jaws are capable of seizing prey larger than it is. As a final coup de grâce it injects chemicals into the victim to paralyse it and start dissolving body tissues, then sucks up the resulting juices.

Status in Aotearoa: Endemic

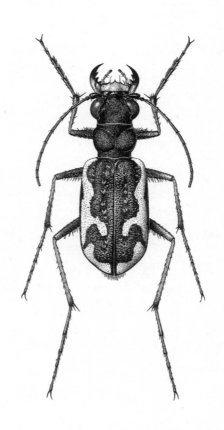

KŪĪ
COMMON TIGER BEETLE

Neocicindela tuberculata

Tiger beetles are as rapaciously predatory as their mammalian namesakes. As entomologist GV Hudson observed, 'They are very voracious, devouring flies, caterpillars and other insects, often considerably larger than themselves.' While clearly less agile, the grub stage is also an effective predator.

Description: Adult beetles are 9–12mm long. Markings on the head and pronotum are dark brown; the elytra are also dark with small green punctures and are bordered with yellow markings. The eyes are large and the mandibles shaped like sickles. Larvae have jaws like those of the adult beetle. Growing to about 20mm long, larvae have special hooks on their backs to help anchor themselves in their burrows.

Habitat and distribution: Found throughout Te Ika-a-Māui North Island and northern Te Waipounamu South Island. Adults are most often seen in the summer months in sunny, open country; deforestation may have benefitted this species. Grubs live in burrows in loose soil or clay banks.

Biology: Both adults and larvae are predators, but with quite different prey capture methods. The long-legged adults are quick but pursue their prey in a stop–start fashion. An overseas study has shown that if tiger beetles run too fast, they don't gather enough photons to form an image of their prey and lose track of it, so need to stop to allow their eyes to catch up. Larvae reside in deep burrows (up to 30cm long), rushing out to seize small prey that gets too close, or allowing larger prey to fall in. The first stage of the burrow is created by the female laying her eggs in the ground; the grub expands and maintains the burrow as it grows. Adult kūī are also capable of rapid flight.

Status in Aotearoa: Endemic

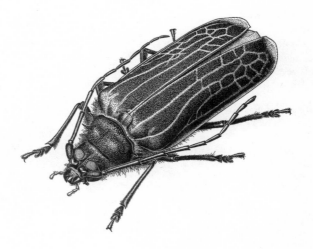

HUHU, TUNGA RERE
HUHU BEETLE

Prionoplus reticularis

This, our largest beetle, can make quite a racket when inadvertently drawn indoors by lights. Their large, plump grubs are regarded as a traditional delicacy.

Description: The beetle stage can reach up to 50mm long and the pronotum is covered in thick brown hairs. The elytra are brown with a network of paler markings. The antennae can be as long as the body. Grubs are plump and pale, reaching up to 75mm in length. The pupal stage looks like an extremely pale version of the beetle.

Habitat and distribution: Grubs are found throughout Aotearoa New Zealand in moist fallen logs, usually softwoods including exotic species such as pine. Beetles start to emerge after pupation from November, with the highest numbers seen between December and February before declining in March.

Biology: Female beetles lay 10–50 eggs in suitable logs. The grub stage is thought to last 2–3 years, with the duration influenced by factors such as temperature and the nutritional quality of the wood. The pupal stage lasts between 3–4 weeks. Beetles do not feed and survive for about two weeks. Grub damage can have impacts on forestry, but these insects have a useful role to play in helping recycle waste logs and other unwanted timber remnants. Beetles are attracted to light, and this may bring them inside, where they can cause quite a commotion as they crash about. They should be handled with care as their jaws can give a painful (but otherwise harmless) bite. Grubs are edible to humans and are a traditional Māori food source; they can be eaten raw or cooked and taste nutty.

Status in Aotearoa: Endemic

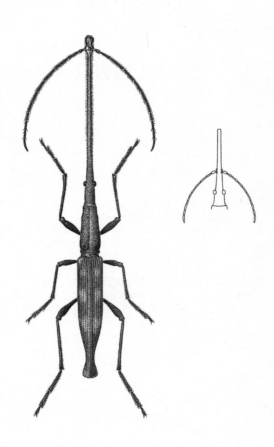

PEPEKE NGUTUROA
GIRAFFE WEEVIL

Lasiorhynchus barbicornis

This species has the longest body length of any Aotearoa New Zealand beetle, thanks to the extremely long rostrum (snout) of males. The position of the antennae differs between males and females, with female antennae positioned away from the mouthparts. This allows females to chew holes and deposit their eggs without hindrance from the antennae. Pepeke nguturoa is just one of many names used by Māori.

Description: Size varies considerably, with larger males reaching close to 90mm long but so-called 'sneaker' males being much smaller. Females are up to 50mm in length. Both sexes are dark brown, with three reddish or yellowish markings on the elytra. Males have a long, heavy rostrum lined with short hairs on the underside and antennae near the tip; females have a shorter, finer rostrum with antennae about halfway along. Grubs may reach a final size of around 40mm before they are ready to pupate.

Habitat and distribution: Forest on Te Ika-a-Māui North Island and north-western Te Waipounamu South Island. Specimens were collected during James Cook's first voyage to Aotearoa (1769).

Biology: The larval stage may last around two years as a wood-boring grub feeding on fungi growing in the tunnels. Pupation occurs in a space near the original entry tunnel, with the beetle chewing its way out to freedom. Sometimes, as noted by entomologist Brenda May, the snouts of stranded beetles may be seen protruding from logs, with the insects 'having apparently failed to make sufficient space to clear their rather wide shoulders'. Adults live several weeks, feeding on sap. Males use their large rostrums in combat with rivals, either biting with their mandibles or trying to push each other off the surface they are on. Being a large and powerful male doesn't always guarantee mating success: 'sneaker' males may sometimes mate with females while larger males are engaged in fighting.

Status in Aotearoa: Endemic

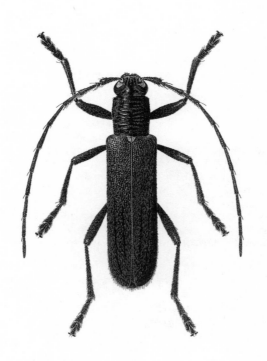

LEMON TREE BORER

Oemona hirta

As the common name suggests, this is a common pest of citrus, although larvae will also feed on a much wider variety of tree and shrub species. Damage caused by larval boring may cause twigs to die.

Description: Beetles range from red-brown to nearly black. Typically a yellow-orange marking can be observed behind each eye and also in the centre of the forward end of the elytra. Beetles range from 15mm to 25mm long. Larval grubs may reach 40mm and are pale with prominent ridges on abdominal body segments.

Habitat and distribution: Found throughout Aotearoa New Zealand, with beetles seen from December to March.

Biology: With its preference for citrus and some ornamentals, this species is regarded as a garden pest. Adults feed on pollen while grubs feed on sapwood. Sufficiently extensive boring by grubs can cause twigs to die and possibly break. Grubs make holes connected to the exterior every few centimetres, allowing the grub to eject debris and providing aeration which makes it difficult for fungi to infest the larval tunnels. Living in tunnels makes it hard to control these grubs with pesticides and provides them with some protection from predators. However, they are not completely safe. The lemon tree borer parasite (*Xanthocryptus novozealandicus*), a handsome black-and-white wasp, can find grubs inside wood. Once located, a female wasp will push her ovipositor through the wood fibres to lay an egg on the lemon tree borer grub. After hatching, the wasp larva proceeds to eat the grub from the inside out.

Status in Aotearoa: Endemic

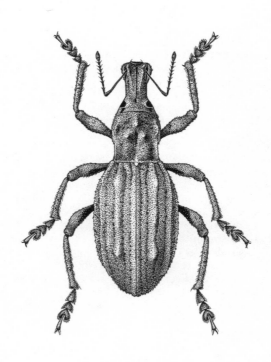

FLAX WEEVIL

Anagotus fairburni

Once thought to be common throughout Aotearoa New Zealand, this weevil has undergone a massive decline in numbers since the arrival of rodents. It is now one of the very few insect species protected under the Wildlife Act 1953. It is illegal to possess, collect or harm these insects without formal authorisation from the Department of Conservation.

Description: Beetles are 20–25mm long, with females slightly larger than males. Colouring is typically in shades of brown due to the presence of slightly iridescent copper- or straw-coloured scales. Older specimens may lose many of these scales, giving them an almost black appearance. There are obvious ridges on the elytra. The larval stage is a cream-coloured grub.

Habitat and distribution: Rodent-free locations only, including alpine areas such as the Tararua Range and offshore islands including Manawatāwhi Three Kings, Mana and several islands in Ata Whenua Fiordland and Te Tauihu-o-te-waka Marlborough Sounds. Because these weevils are associated with harakeke New Zealand flax (*Phormium tenax*) and wharariki mountain flax (*P. cookianum*), their range was probably much greater before the introduction of rodents to Aotearoa.

Biology: Adults may look sturdily built, but they are flightless and easy prey for rodent predators. Both adults and larvae are found on flax plants, with adults hiding by day to emerge at night to feed on the leaves. Feeding damage is typically in the form of irregular, rough-edged notches on the leaf margins. Larvae feed on the root system. In the absence of rodents in particular, these weevils can flourish. A population of eighty adults introduced to Mana Island near Te Whanganui-a-Tara Wellington in 2004 has grown to tens of thousands, with the flax plants at the initial release site having effectively been eaten to extinction.

Status in Aotearoa: Endemic

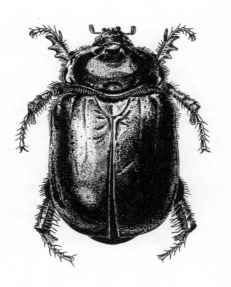

NGUNGUTAWA
LARGE SAND SCARAB BEETLE

Pericoptus truncatus

This is the largest scarab beetle species in Aotearoa New Zealand. The larvae are unique among sand scarabs globally in that they have broad oar-like claws, which may allow them to move more easily in sandy habitat. Despite being creatures of the land, larvae have been observed submerging themselves in sea water for short periods.

Description: Grubs are pale and C-shaped, reaching up to 65mm in length. Their limbs feature uniquely modified claws that might help them navigate sandy conditions more easily. Beetles are up to 30mm long. The upper side is glossy brown-black, while the underside is coated in yellow hairs.

Habitat and distribution: Associated with driftwood on sandy beaches throughout Aotearoa. Adults are seen from March to May; they overwinter in the sand and appear again to mate and lay eggs from September to November.

Biology: Larvae are known to feed on rotting driftwood and roots of dune grasses. They have rather exacting habitat moisture requirements and have been recorded moving as much as 80m at night to find conditions that are neither too wet nor too dry. Entomologist GV Hudson noted them as being particularly strong, and able to push the lid off a container to escape. They occasionally submerge themselves in sea water, possibly to remove external parasites. The larval lifecycle may take 2–5 years before pupation. Beetles appear in autumn, but bury themselves 60–120cm deep under the sand for the winter months before re-emerging to reproduce in the spring. The beetle stage is not known to feed at all. Shore birds such as variable oystercatchers (*Haematopus unicolor*) have been recorded eating sand scarab larvae. The introduced yellow flower wasp (*Radumeris tasmaniensis*) paralyses sand scarab larvae to serve as food for their grubs.

Status in Aotearoa: Endemic

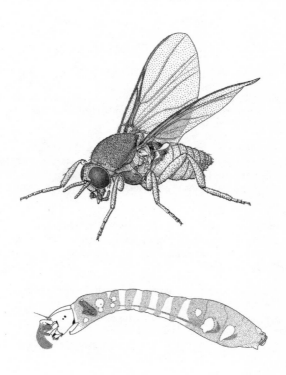

NAMU
NEW ZEALAND BLACKFLY, SANDFLY

Austrosimulium sp.

In Aotearoa New Zealand these are called sandflies or namu, but worldwide the family Simuliidae are known as blackflies. The genus *Austrosimulium* is only known from Aotearoa, Tasmania and mainland Australia. There are nineteen austrosimulium species living in Aotearoa, and only three of these bite humans. All the others are bird-biters. Species from overseas transmit diseases, but in Aotearoa only bird-diseases are known.

Description: The three human-biting species are very similar in size (2–3mm long) and it is difficult to tell them apart without a microscope. The *ungulatum* species-group has a basal tooth on the tarsal claws of the female; this is absent in the *australense* species-group.

Habitat and distribution: Found near water and in humid bush, for example at beaches, lakes, rivers and swamps. *Austrosimulium australense* occurs with *A. tillyardianum* in both main islands, but can be found further north up to Te Rerenga Wairua Cape Reinga and south to Rakiura Stewart Island, while *A. tillyardianum* is more abundant along the east coast. *A. ungulatum* can be found in Te Tai Poutini Westland and Ata Whenua Fiordland.

Biology: Only the females bite, as they require blood to produce large quantities of eggs. Namu copulate only once in a lifetime; females store sperm to fertilise later batches of eggs, laid in running water. The larvae (below) hatch in about ten days and attach themselves to vegetation or rocks in flowing water, using a set of fans to filter for food. They form a cocoon of silk in which the larvae develop into adult flies (above). These bite by piercing the skin with their knife-shaped mouthparts, releasing a powerful histamine that prevents the blood from clotting and agglutinins that prepare the pooled blood for digestion. This mix causes the red bumpy reaction in the victim's skin.

Status in Aotearoa: Endemic

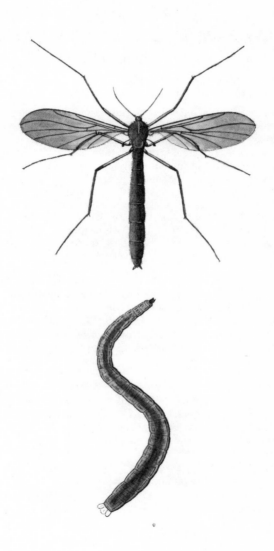

TITIWAI
NEW ZEALAND GLOW-WORM

Arachnocampa luminosa

The Māori name, titiwai, refers to lights reflected in water. The genus name *Arachnocampa* means spider-worm, for the way the larvae hang sticky silk threads to catch small invertebrates. Larvae glow to attract prey – mainly small aquatic insects – into their threads; the bioluminescence is the result of a chemical reaction, which has been known from beetles in Europe. Entomologist GV Hudson had a hard time convincing scientists that this 'worm' was the larva of a fly (a fungus gnat), but finally succeeded after rearing experiments.

Description: Adults (above) are around 16mm. Only 3–5mm long when it emerges from the egg, the larva (below) grows to about 3cm through several instars. Its pale body is soft while the head capsule is hard.

Habitat and distribution: Glow-worms need moisture; they can be found throughout Aotearoa New Zealand along streambeds in the bush or in caves.

Biology: This species spends most of its life (6–12 months) in its larval form, the time depending on available food. When it outgrows the hard head capsule the larva moults. It hangs down as many as seventy threads of silk (called snares), each holding droplets of mucus. In caves the snares are up to 40cm long but in forests only up to 5cm, due to wind entangling longer snares. Once prey is caught in a snare, the larva pulls it up by ingesting the snare and starts feeding. Cannibalism occurs when population densities are high or when adult flies entangle themselves in snares after hatching. A cave species of harvestman is known to prey on glow-worms in some caves, and a fungus will also gradually kill larvae; otherwise they have few predators.

Status in Aotearoa: Endemic

BAT-WINGED FLY

Exsul singularis

Considered one of the rarest and most unusual flies in the world, this insect was first recorded in 1901 and until the late 1980s only a handful of specimens was known.

Description: A rather strange-looking insect with a black body. Males (shown here) have a pair of large, dark, rounded wings, almost like those of a butterfly, while female wings are more like those of a house fly.

Habitat and distribution: First discovered in Piopiotahi Milford Sound in 1901, it has since been found as far north as Aoraki National Park. Most specimens have been collected at altitudes over 1000m; the relative inaccessibility of their habitat may have been a factor in their perceived rarity.

Biology: Though not as rare as once thought, the bat-winged fly remains extremely under-researched and little is known about its life history. Records show that larvae need two years to develop and are carnivores. Sometimes they can be found gathering on rocks in groups of up to one hundred individuals. Like other alpine insects, they bask in the sun. It is likely that the flies absorb heat from the sunlight, using their big wings like solar panels. Their dark colouring is typical of insects from the alpine zone, as it allows them to soak up the maximum amount of warmth from the sun to combat changeable, often frigid temperatures. Another explanation for the males having large, prominent wings could be competition; conspicuous items such as these are a good way of attracting females while not hindering movement.

Status in Aotearoa: Endemic

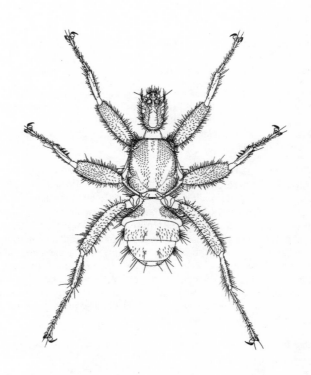

BAT FLY

Mystacinobia zelandica

This blind, wingless fly is entirely dependent on a supply of guano and the warmth of a bat roost to survive. First discovered on a bat in the 1950s, it wasn't until Kopi, a famous kauri (*Agathis australis*) in Omahuta, Northland, collapsed under its own weight in 1973 that the bat fly was formally described. A few specimens were found on a short-tailed bat that had been killed when Kopi fell, and were sent to Beverley Holloway, an entomologist working at the Department of Scientific and Industrial Research (DSIR) who gained her PhD in biology at Harvard and was awarded the New Zealand Commemoration Medal.

Description: This species is 4–9mm long with specially adapted long bristly claws which help with movement in bat fur. Males are larger than females and have hairier thoraxes, with females having larger and more distended abdomens.

Habitat and distribution: In the wild, the bat fly lives only in tree roosts of the nationally threatened endemic short-tailed bat (*Mystacina tuberculata*) in a very few places across Aotearoa New Zealand, but it is believed that its ancestors evolved in Australia.

Biology: There can be thousands of bat flies of all life stages from egg to adult in a bat roost inside a tree hollow, living on the guano-coated walls. Some adults, mostly gravid females, will cling to bats, potentially the founders of a new colony should the bats abandon a fallen tree. Research suggests that these insects need temperatures approaching 30°C to prosper; without bats to generate heat inside the enclosed space of a roost, individual bat flies will soon die. Males live longer than females. Older males that have already mated may produce sound, possibly to deter bats from feeding on the flies – the bats normally consume insects, as well as pollen, which is cleaned and consumed by the flies while grooming the fur of the bats. Related species overseas are parasitic on the bats they live with, while the endemic species lives in a commensal relationship with the bats.

Status in Aotearoa: Endemic

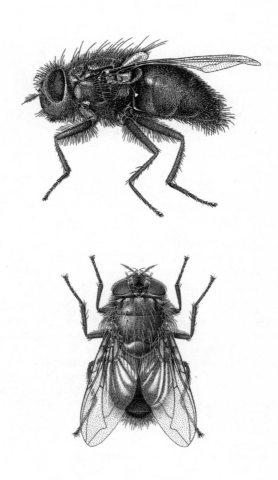

RANGO PANGO
NEW ZEALAND BLUE BLOWFLY

Calliphora quadrimaculata

Our largest native blowfly, and also a large blowfly in world terms. Unlike most blowfly maggots that generally feed on animal tissue or faeces, this species can utilise decaying plant tissue, such as tussock, as well as animal tissue.

Description: Adults have a body length of 9.5–15mm. The eyes are densely haired. In males the eyes meet on the mid front line of the head, while in females they are separated. The most distinctive feature is the very large orange spiracles (breathing openings) on the thorax. The thorax is black, with the middle part of the back evenly grey-dusted and the lower part a brownish colour. The legs have a blackish brown femur with a thin grey dusting; the tibiae are a reddish brown. The abdomen is black with stunning metallic royal blue reflections.

Habitat and distribution: Found throughout Aotearoa New Zealand including more remote island groups such as Rēkohu Chatham, Motu Maha Auckland and Motu Ihupuku Campbell Islands. It can survive in a range of habitats, including areas of snow tussock over 1000m in altitude.

Biology: The lifecycle from egg to adult takes around three weeks, with warmth accelerating development. Eggs hatch around a day after being laid. The larval phase (three stages) lasts a little over a week before pupation, and the adult fly emerges about two weeks later. Adults typically live for 2–3 weeks. This species is not a pest. Although they can transfer bacterial diseases between animals, including humans, they are also pollinators and their larvae have an important role in clearing up decaying biological material.

Status in Aotearoa: Endemic

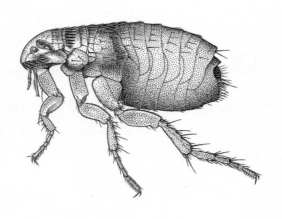

BIRD FLEAS

Parapsyllus spp.

Half of the flea species found in Aotearoa New Zealand are native, with the only endemic mammal flea being the bat flea *Porribius pacificus*. Endemic bird fleas are mainly associated with seabirds. Only *Parapsyllus nestori* is parasitic on land birds (such as kea and parakeets). Interestingly, Apterygidae (kiwi) are normally not parasitised by fleas.

Description: The white, ovoid eggs are microscopic. Larvae are about 2mm long and have a grub-like appearance. The wingless adults are 1–2mm long, usually a reddish-brown colour and laterally compressed. This allows them to slip between dense hairs and feathers on the host just above the top layer of the skin. Different from a cat flea (shown here), *Parapsyllus* lack the comb-like spines below and at the back of the head.

Habitat and distribution: Mainly found on ground-nesting birds. The last Ice Age, which occurred between 15,000 and 100,000 years ago, must have forced kea down the alps and shoreward until they had to nest under large boulders and among rocks close to tītī muttonbirds, and thus obtained their fleas. After the end of the Ice Age, when kea retreated to the mountains and lost contact with the seabirds, these fleas could subsequently have evolved into *P. jacksoni* on tītī and into *P. nestori* on kea.

Biology: Adult fleas starve within days if separated from their warm-blooded host and they must feed on blood to become capable of reproduction. A gravid, or pregnant, female will then produce eggs every day until she dies. The larvae's most important food is crumbs of dried blood excreted by adult fleas as faecal material. Larvae require moisture and warmth and hide in the nesting substrate to avoid light. After four stages the larvae spin an adhesive cocoon, which is quickly camouflaged by dirt and dust. Emerging adults detect variations in light, temperature and carbon dioxide in the presence of a potential host and use their legs (adapted with a pad of elastic protein) to jump onto it.

Status in Aotearoa: Endemic

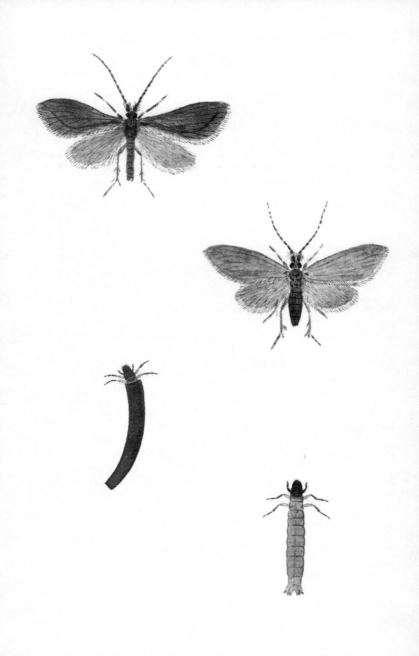

SMOOTH-CASED CADDIS

Olinga feredayi

Most caddisfly larvae construct a case structure (below left) to protect and disguise the larval stage while still enabling them to remain mobile. Different groups of caddisflies can be distinguished by the different shapes and materials they use for their cases. *Olinga* caddis larvae build smooth cases that lack sand grains. The aperture of the case is straight to slightly curved, resembling a horn, and indeed it is also known as horn-cased caddis. This species is a very good water-quality indicator for hard-bottom streams.

Description: The rounded orange or red head of the larva (below right) is smooth and without hairs. There is a black mark on the side of the body behind the hind legs. Adults (male above left, female above right) are rather small with yellow and brown, very fringed wings.

Habitat and distribution: Common, and generally distributed throughout Aotearoa New Zealand in and around running streams. Larvae are most common in bush-covered, cold-water, stony streams.

Biology: The very active larvae climb on rocks in strong water currents. Before they pupate, inside their cases, they attach themselves to a rock, often in large groups, and close the upper end of the case. Adults are active in the evening at dusk, often being seen in swarms as they emerge simultaneously. Females carry their eggs at the end of their body before dispatching them in the water. An abundance of *Olinga* larvae indicates good habitat and water quality, particularly if mayfly and stonefly nymphs are also present.

Status in Aotearoa: Endemic

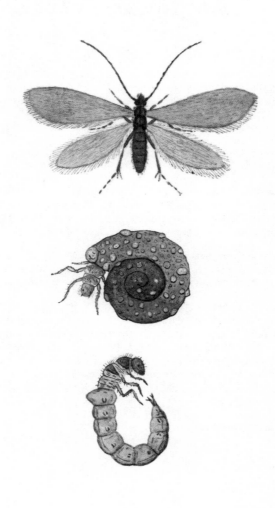

SPIRAL-CASED CADDIS

Helicopsyche spp.

These larvae construct a small mobile case made of sand grains arranged in a spiral (centre). At first glance they look like a tiny snail, but their legs can be seen when the larva moves. They even fooled early naturalists, who described them as actual shells.

Description: The wings of adults (above) are fringed with long hairs, and dusky black with a pale spot on the forewings, which are slightly transparent. The whorls of the larva's case are flattened; the inner is very smooth because miniature grains are used here. The long larva (below) is curved with a large oval reddish head. The first part of the thorax, often exposed from the case, is hard, shiny and reddish, too, while the parts below are softer and rather green. The thorax carries some single bristles, sitting on tubercles at the third segment. The abdomen appears polished and its first segment is humped. The front legs are shorter than the others, and the end segment carries claws, typical for caddisflies. The pupae are green and brown, with yellowish eyes and already green wings, curved along the body in the same way the antennae and legs do.

Habitat and distribution: Larvae are common in many stony or gravelly, bush-covered streams throughout Aotearoa New Zealand.

Biology: The presence of larvae indicates good habitat and water quality, especially if they occur together with mayflies or stoneflies. Despite the larvae being around all year and being especially abundant in mid-winter, in clusters of 50–100, they are easy to overlook. When older and larger, they become more active. They pupate in the sealed case, although pupae will leave the case for a swim before the adult emerges in December or January.

Status in Aotearoa: Endemic

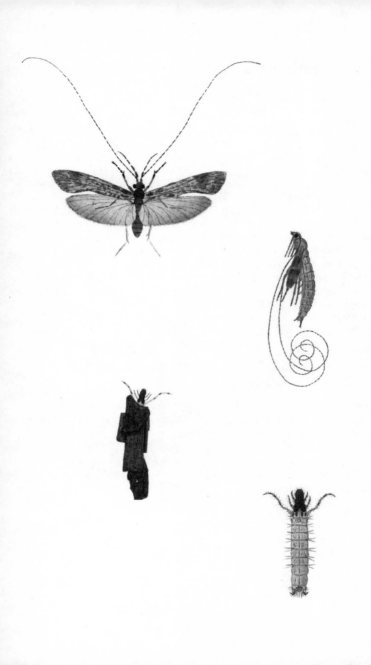

STICK CADDIS

Triplectides obsoleta

Instead of constructing a case from particles of different types, the larvae of this species use a hollowed piece of stick (below left), boring into the core of a suitable piece. These are normally simple, straight and much longer than the larvae so that they can retreat into the middle and be completely safe. Sometimes more hollowed plant material is added. Among debris washing down in a stream, these cases are practically invisible.

Description: Adults (above left) have very long antennae and hairy mouthparts. The wings are grey with white and black dots and stripes. The legs are pale, with the front legs half black. The larvae (below right) have long hind legs with distinct dark bands between joints. Seven pairs of respiratory appendices are visible on the abdomen. The last abdominal segments carry bristles and small hooks.

Habitat and distribution: Larvae are common in hard-bottom and soft-bottom streams, in areas of bush cover, farmland and sometimes urban land, throughout the year but peaking in midsummer. Found throughout Aotearoa New Zealand.

Biology: Larvae can occur in streams with moderate pollution, so are not necessarily an indicator of good water quality. They feed on plant material that falls into the stream. The longer middle and hind legs are used for locomotion, feeding or holding on to a surface. Prior to pupation they spin silk to attach their case to a submerged surface such as a log, close both ends of the case and form a cocoon. Pupae (above right) have the enormously long antennae of adults, curled up at the ends. After leaving their case they are washed in strong rapids downstream, where the adults emerge during a long period between November and March. Adults are active at dusk, displaying a typical zigzag flying pattern.

Status in Aotearoa: Endemic

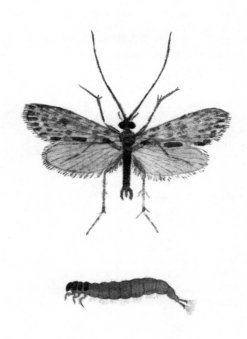

NET-SPINNING CADDISFLY

Aoteapsyche colonica

This species constructs filter-feeding nets out of silk to trap drifting particulate food items, such as detritus, algae and even other invertebrates. Each silk thread is placed to create mesh webbing that stretches and balloons with food in the water current. A construction of this kind is quite rare to find in the animal kingdom. The best-known organism to do this is of course the spider.

Description: In adults (above), the small body appears blackish and hairy, with pale strips along both sides. The antennae are ringed and the legs yellow. The wings are narrow and greyish with dark markings. Larvae (below) have a narrow, hard and darker front with two short front legs, typical for caddis, and have four plates and a hump on their backs. The abdomen is pale, soft and much wider than the front. This species has characteristic gill tufts along the belly.

Habitat and distribution: Larvae are abundant in many stony streams and rivers, throughout Aotearoa New Zealand, reaching greatest abundance downstream of lake outlets.

Biology: Larvae are common in streams of moderate to good quality, so are not particularly useful as water-quality indicators. They have less tolerance in hard-bottom streams than in soft-bottom streams though. They prefer very strong rapids that can wash the food into their nets. The nets are attached to their loosely-built shelter, which sits tightly on a rock. The nets are really difficult to spot under water but become visible when they dry. Adults are often seen sitting along the streambanks. Larvae can be found all year round but are most abundant in early spring.

Status in Aotearoa: Endemic

PEPE TUNA
PŪRIRI MOTH

Aenetus virescens

Our largest species of native moth, they can, as the English common name suggests, be associated with pūriri trees, but are known to attack around seventy other kinds including many introduced species. The Māori name pepe tuna refers to the use of the caterpillars as bait for eels.

Description: Female moths have a wingspan approaching 150mm; male wingspans are closer to 100mm across. They are typically patterned in shades of green, but other colours including red, yellow and even albino forms have been recorded. Male hindwings are typically much paler than those of females. Caterpillars may reach 100mm long and are purplish-pink with dark-brown heads.

Habitat and distribution: Found on Te Ika-a-Māui North Island only, with moths usually seen in late spring to early summer. Larvae may be found at any time of year by locating the distinctive camouflaged webbing that covers their feeding scars on tree hosts.

Biology: After hatching from eggs laid on the forest floor, young caterpillars feed on fungi before transferring to trees where they will spend the bulk of their lives. Pepe tuna caterpillars create a burrow shaped like a 7 inside the tree, covering the entrance with tough silk webbing which makes a very effective camouflage. They feed on callus tissue around the tunnel entrance. The caterpillar stage may span five years, while the pupal stage is thought to last about three months. The moths are very short-lived, lasting around a week or so. They lack mouthparts so do not feed. After the moths have emerged, abandoned tunnels are often utilised by other insects such as tree wētā (*Hemideina* spp.).

Status in Aotearoa: Endemic

ĀWHETO
VEGETABLE CATERPILLAR AND
FOREST GHOST MOTH

Ophiocordyceps robertsii and *Aoraia dinodes*

The vegetable caterpillar or āwheto (āwhato) is a fungus that infects the ground-dwelling caterpillars of the forest ghost moth, ultimately turning them into mummified husks inside a fungal casing. Māori dried and burned āwheto, mixing the charcoal with muttonbird fat to make an antiseptic ink used in moko (tattooing). Āwheto were popular as a curiosity among Pākehā in the early 1900s.

Description: Adult moths (above) have a wingspan up to 70mm across, with intricate whitish forewing markings on a background in shades of brown. The hindwings and abdomen are fawn to brown. The mature āwheto (below) looks like a dry, leathery version of a caterpillar with one or more stalks growing upwards from the head end.

Habitat and distribution: The moth is found in cool temperate and subalpine native forest in southern Te Waipounamu South Island and on Rakiura Stewart Island. First collected near Waihōpai Invercargill, it may now be locally extinct there. Adults are most likely to be seen in March and April. Āwheto occurs on both main islands and infects several moth species.

Biology: The lifecycle of the forest ghost moth may take 2–3 years. Eggs are laid in leaf litter. Caterpillars crawl through the litter, feeding on fallen leaves and turf, and may be unlucky enough to also consume spores of āwheto fungus. These germinate inside the insect, and when the caterpillar goes underground to pupate, the āwheto's development picks up speed. It consumes the caterpillar while using its body like scaffolding as it spreads its hyphae throughout, ultimately resulting in a hard, caterpillar-shaped object with stalks arising from the head. These stalks extend above ground and release spores to infect new caterpillar hosts.

Status in Aotearoa: Endemic (both species)

NEW ZEALAND TIGER MOTH

Metacrias erichrysa

This moth species has brightly coloured males, while females are dull with vestigial wings and never leave the pupal cocoon. After the female lays her eggs in the cocoon, she dies and her corpse serves as food for her offspring.

Description: Males have black bodies. They have strongly contrasting black and yellow-orange wing markings, with a wingspan typically 37mm across. The almost wingless females are broadly oval when viewed from above and are covered in matted white-ochre hairs. Caterpillars are coated with spine-like hairs and are coloured in alternating bands of golden yellow and black. The black bands are dotted with iridescent blue spots.

Habitat and distribution: Subalpine herb fields and shrubs in open country in wetter parts of mountain ranges in both Te Waipounamu South Island and lower Te Ika-a-Māui North Island. Moths are usually seen between 900 and 1200m above sea level although higher-altitude records are known. Males are usually on the wing in January.

Biology: Caterpillars feed on a variety of species including snow tussock (*Chionochloa rigida*) and various species of *Raoulia*, *Muehlenbeckia* and *Senecio*. Larvae are present from midsummer and undergo winter diapause. Pupae are covered by a rather flimsy cocoon which the flightless female never leaves. Males can locate females by homing in on their pheromones. After mating, the females lay their eggs inside their cocoons. An informally published observation by lepidopterist Brian Patrick noted that the corpses of females also serve as food for their young. Males are strong fliers, but the immobility of females means that dispersal to new areas can only be accomplished through migration of larvae.

Status in Aotearoa: Endemic

NORTH ISLAND LICHEN MOTH

Declana atronivea

This moth is remarkable for its striking black-and-white colouring and for being one of only two insect species known to have asymmetric wing patterns. It is a master of disguise as both a caterpillar and a moth.

Description: Moths (above) have a wingspan of up to 50mm; the striking black-and-white colour pattern differs between facing forewings. Hindwings are grey-brown. The South Island lichen moth (*Declana egregia*) is similar but has less-elaborate, symmetrical wing markings. Caterpillars are white and brown and resemble twigs of puahou fivefinger (*Pseudopanax arboreus*) or even bird droppings.

Habitat and distribution: Found only on Te Ika-a-Māui North Island. Moths appear to emerge in two waves, one in early spring and another in late summer. Caterpillars may be found on their host plants (several species of *Pseudopanax*).

Biology: Caterpillars hatch around early December and pupate some 5–6 weeks later, with moths emerging after a few more weeks. Sometimes the pupal stage (below) may overwinter with moths appearing in spring. At rest on a plain background the moth is extremely easy to see, but when situated on an irregular background such as lichen it is much more difficult to spot. The wing pattern asymmetry helps break up the outline of the moth's shape, making it less obvious to predators such as birds. Caterpillars are also adept at deception. When curled up, the position of the various areas of white or brown gives them the appearance of bird droppings; when anchored with the body projected erectly, they can look like a twig carrying spots of lichen. The pupal stage is also camouflaged under a thin cocoon covered in pellets of earth or dead leaves.

Status in Aotearoa: Endemic

KAHUKURA
RED ADMIRAL BUTTERFLY

Vanessa gonerilla

Perhaps our most iconic endemic butterfly, the kahukura is reportedly declining in numbers, possibly due to the reduced availability of preferred food plants (particularly in urban areas) and being targeted by several exotic species of wasp. The Māori name kahukura means red cloak.

Description: Wingspan is 50–60mm across. The upper surfaces of the forewings have a brown area close to the body and are black further away. They also have prominent red areas, and this distinguishes them from the superficially similar yellow admiral (*Vanessa itea*), which has yellow markings instead. The red areas of the hindwing contain several black-bordered spots. Kahukura caterpillars, which grow to about 36mm, are generally dark with some light striping; some may be pale. Spiny tufts are dotted across the upper surface of the body.

Habitat and distribution: Rarely seen in Tāmaki Makaurau Auckland, possibly due to the declining availability of ongaonga giant nettle (*Urtica ferox*), the caterpillar's preferred food plant; more common further south. Butterflies are usually seen in summer but can overwinter if nectar-producing flowers are available. A subspecies (*V. g. ida*) inhabits Rēkohu Chatham Islands.

Biology: Although preferring ongaonga, caterpillars will also feed on other nettle species. Caterpillars bend over leaf tips and fix them in place with silk to provide shelter, and possibly to hide from predators such as birds; caterpillars are also targeted as prey by several species of wasp. There can be several generations over the summer. The time between emergence from the egg and pupation is about six weeks, while the pupal stage lasts 14–18 days. Butterflies may overwinter, generally sitting dormant unless conditions are warm and sunny enough for flight. They are active fliers rather than gliders.

Status in Aotearoa: Endemic

PEPE POURI
BLACK MOUNTAIN RINGLET

Percnodaimon merula

As an alpine specialist, this butterfly could literally be said to be living the high life. Despite its seeming fragility, this species has a number of adaptations that help it survive in the challenging and dynamic environment of mountain scree slopes.

Description: Butterflies are dark brown to almost black with a series of black-bordered pale eyespots on the forewings. Maximum wingspan is about 55mm, with females slightly larger than males. Caterpillars can be blue grey to pale brown and attain a maximum length of around 25mm.

Habitat and distribution: Found in alpine regions of Te Waipounamu South Island between 800m and 2500m altitude. Butterflies are usually seen flying over rocks on sunny days from November to March.

Biology: This species might be deemed a solar-powered butterfly. A combination of dark colouring and sun-seeking behaviour allows it to make the most of any warm sunshine. Eggs are laid on rocks rather than food plants as rocks heat up more and hold heat for longer, making better incubators. Even the pupal stage has a heat-seeking strategy. Most butterfly pupae hang vertically, but pepe pouri pupae are tucked into horizontal crevices to take advantage of any rock warming, which will help speed up development. Caterpillars trade warmth for security, emerging at night to avoid predators as they feed on the tips of blue tussock grass (*Poa colensoi*). They may take two or three summers to become ready to undergo metamorphosis and emerge as butterflies.

Status in Aotearoa: Endemic

GLOSSARY
REFERENCES
ACKNOWLEDGEMENTS
IMAGE CREDITS

GLOSSARY

Abdomen The rearmost of the three major body areas in insects (head, thorax and abdomen). It contains the structures necessary for respiration, reproduction, digestion and excretion.

Antennae Colloquially called feelers, these are a pair of sensory organs on the heads of insects. They are used to sense touch, smell and other stimuli.

Carnivore An animal that eats animal matter.

Cerci A pair of appendages on the rear of many kinds of insect. These have a variety of forms and functions, e.g. hard, curved structures used as weapons by earwigs, or slender filaments used as sensors in stonefly nymphs.

Class The taxonomic rank between phylum and order. Insects belong to class Insecta.

Commensal An organism that derives benefits (e.g. food) from a close association with a second species that is not helped or harmed by this arrangement.

Detritus Decomposing particulate organic material.

Diapause A period of delayed development in response to unfavourable environmental conditions.

Dun A colloquial name for the subimago stage.

Femur The first long section of the insect leg. It is attached to the body by two smaller parts (coxa and trochanter) and is followed by the tibia.

Elytra Forewings that have lost their flight function and have become protective wing cases for the hindwings.

Genus A taxonomic rank between family and species. In a species combination, the genus name goes first, e.g. *Hemideina* is the genus name in *Hemideina crassidens*.

Gills Structures in aquatic insect larvae that allow the exchange of oxygen and carbon dioxide to occur while the insect is submerged.

Gravid Full of well-developed eggs.

Haltares Small, often drumstick-like structures that replace the hindwings in flies. These help with flight stability and orientation.

Herbivore An animal that eats plant matter.

Imago The adult stage of an insect. Except for mayflies, which have a subimago or dun stage, winged insects are all adults.

Instar The insect form between two periods of moulting, or just after hatching from an egg (first instar).

Larva The immature form of insects that have a pupal stage.

Mandibles Insect mouthparts, colloquially referred to as jaws. Depending on their shape, they may be used for biting, fighting, holding objects or a variety of other purposes.

Naiad The name for dragonfly and damselfly nymphs.

Nymph The immature form of insects that do not have a pupal stage.

Omnivore An animal able to eat both plant and animal matter.

Ootheca An egg case containing multiple eggs made by some insects, most commonly mantids and cockroaches.

Order The taxonomic rank between class and family, e.g. order Orthoptera.

Ovipositor A tube-shaped organ used to lay eggs.

Parasite An organism that draws nutrients from (and often lives on or in) another organism (the host). Parasites may cause harm to their hosts, but do not usually kill them (e.g. feather lice).

Parasitoid In insects this is a larval stage that draws nutrients from and develops on or in another organism (the host), eventually killing it (e.g. larvae of spider-hunting wasps).

Parthenogenesis Reproduction without the involvement of males.

Phylum The taxonomic rank between kingdom and class. Insects belong to phylum Arthropoda.

Pronotum The upper surface of the first segment (prothorax) of the thorax.

Rostrum A beak-like projection on the head of an insect that ends with the mouthparts.

Soldier (termite) One of four termite castes, or specialised forms, soldiers are larger than the more common workers and have disproportionately larger heads and mandibles. They guard the termite colony.

Spiracle An opening in the wall of the abdomen to allow air to enter the tracheal system, which distributes oxygen around the insect body.

Stage The major divisions in the insect lifecycle (egg, larva, pupa, adult).

Stylet A slender, hollow feeding tube used by some insects such as aphids.

Subimago The sexually immature winged life stage between nymph and adulthood that is unique to mayflies.

Tarsus The leg segment furthest from the body and connected to the tibia. It carries the insect's claws.

Thorax The second of three main body areas in an insect. It consists of three segments called (from front to back) the prothorax, the mesothorax and the metathorax. Each of these segments bears one pair of legs, while wings are attached to the meso- and/or metathorax.

Tibia The leg segment between the femur and the tarsus.

REFERENCES

INTRODUCTION

Buckley, TR, NP Lord, A Ramón-Laca, JS Allwood and RAB Leschen, 'Multiple lineages of hyper-diverse Zopheridae beetles survived the New Zealand Oligocene Drowning', *Journal of Biogeography*, vol. 47, no. 4, 2020, pp. 1–14, doi: 10.1111/jbi.13776.

Macfarlane, RP, PA Maddison, IA Andrew, JA Berry, P Johns, RJB Hoare, M-C Larivière, P Greenslade, RC Henderson, CN Smithers, R Palma, JB Ward, RLC Pilgrim, RAB Leschen, D Towns, I McLellan, D Teulon, JF Lawrence, G Kuschel, D Burckhardt, TR Buckley and SA Trewick, 'Phylum Arthropoda. Subphylum Hexapoda. Protura, springtails, Diplura, and insects', in DP Gordon (ed.), *New Zealand Inventory of Biodiversity*, vol. 2, Canterbury University Press, Christchurch, 2010, Chap. 9, pp. 233–467.

Wharton, DA, 'Cold tolerance of New Zealand alpine insects', *Journal of Insect Physiology*, vol. 57, no. 8, 2011, pp. 1090–1095, doi: 10.1016/j.jinsphys.2011.03.004.

Winterbourn, MJ, 'The faunas of thermal waters in New Zealand', *Tuatara*, vol. 16, no. 2, 1968, pp. 111–121.

FOR SPECIFIC SPECIES

Arensburger, P, C Simon and K Holsinger, 'Evolution and phylogeny of the New Zealand cicada genus *Kikihia* Dugdale (Homoptera: Auchenorrhyncha: Cicadidae) with reference to the origin of the Kermadec and Norfolk Islands' species', *Journal of Biogeography*, vol. 31, no. 11, 2004, pp. 1769–1783, doi: 10.1111/j.1365-2699.2004.01098.x. [*Kikihia* spp.]

Berry, JA, 'Alysiinae (Insecta: Hymenoptera: Braconidae)', *Fauna of New Zealand*, no. 58, Manaaki Whenua Press, Lincoln, 2006, pp. 1–95. [*Aphaereta aotea*]

Brookes, AE, 'A new genus and six new species of Coleoptera', *Transactions and Proceedings of the New Zealand Institute*, vol. 63, 1932, pp. 25–33. [*Anagotus fairburni*]

Brown, WL, 'A review of the ants of New Zealand (Hymenoptera)', *Acta Hymenopteralogica*, vol. 1, no. 1, pp. 1–50. [*Huberia striata*]

Chapman, RB, 'Pasture pests', in RR Scott (ed.), *New Zealand Pests and Beneficial Insects*, Lincoln University of Agriculture, Lincoln, 1984. [*Teleogryllus commodus*]

Craig, DA, REG Craig and TK Crosby, 'Simuliidae (Insecta: Diptera)', *Fauna of New Zealand*, no. 68, Manaaki Whenua Press, Lincoln, 2012, pp. 1–336. [*Austrosimulium australense*]

Dear, JP, 'Calliphoridae (Insecta: Diptera)', *Fauna of New Zealand*, no. 8, Department of Scientific and Industrial Research, Wellington, 1985, pp. 1–88. **[*Calliphora quadrimaculata*]**

Dugdale, JS, 'Hepialidae (Insecta: Lepidoptera)', *Fauna of New Zealand*, no. 30, Manaaki Whenua Landcare Research, Lincoln, 1994, pp. 1–164. **[*Aoraia dinodes*; *Ophiocordyceps robertsii*]**

Forster, RR and LM Forster, *Small Land Animals of New Zealand*, John McIndoe Ltd, Dunedin, 1975, p. 175. **[*Huberia striata*]**

Gibbs, G, *Weta, New Zealand Wild series*, Reed Publishing, Auckland, 2003. **[*Hemideina crassidens*]**

Gibbs, GW, 'The New Zealand genus *Metacrias* Meyrick (Lepidoptera: Arctiidae) systematics and distribution', *Transactions of the Royal Society of New Zealand: Zoology*, vol. 2, no. 19, 1962, pp. 153–167. **[*Metacrias erichrysa*]**

Gibbs, GW, *The New Zealand Butterflies: Identification and natural history*, Collins, Auckland, 1980. **[*Vanessa gonerilla*]**

Gibbs, GW, 'Habitats and biogeography of New Zealand's deinacridine and tusked weta species', in LH Field (ed.), *The Biology of Wetas, King Crickets and their Allies*, CABI Publishing, Wallingford, 2001, pp. 35–55. **[*Deinacrida rugosa*]**

Gilbert, C, 'Visual control of cursorial prey pursuit by tiger beetles (Cicindelidae)', *Journal of Comparative Physiology A*, vol. 181, 1997, pp. 217–230. **[*Neocicindela tuberculata*]**

Giles, ET, 'The biology of *Anisolabis littorea* (White) (Dermaptera: Labiduridae)', *Transactions and Proceedings of the Royal Society of New Zealand*, vol. 80, 1952, p. 383. **[*Anisolabis littorea*]**

Harris, AC, 'Pompilidae (Insecta: Hymenoptera)', *Fauna of New Zealand*, no. 12, Manaaki Whenua Press, Lincoln, 1987, pp. 1–154. **[*Sphictostethus fugax*]**

Hitchings, TR and AH Staniczek, 'Nesameletidae (Insecta: Ephemeroptera)', *Fauna of New Zealand*, no. 46, Manaaki Whenua Press, Lincoln, 2003, pp. 1–72. **[*Ameletopsis perscitus*; *Aoteapsyche colonica*; *Helicopsyche zealandica*; *Nesameletus ornatus*; *Olinga feredayi*; *Triplectides obsoleta*; *Zephlebia dentata*]**

Holloway, BA, 'A new bat-fly family from New Zealand (Diptera: Mystacinobiidae)', *New Zealand Journal of Zoology*, vol. 3, no. 4, 1976, pp. 279–301. **[*Mystacinobia zelandica*]**

Hudson, GV, *New Zealand Moths and Butterflies*, West, Newman & Co., London, 1898, pp. 1–144. **[*Declana atronivea*]**

Hudson, GV, *New Zealand Neuroptera : A popular introduction to the life-histories and habits of may-flies, dragon-flies, caddis-flies and allied insects inhabiting New Zealand, including notes on their relation to angling*, West, Newman & Co., London, 1904. **[*Ameletopsis perscitus*; *Aoteapsyche* *colonica*; *Archichauliodes diversus*; *Helicopsyche zealandica*; *Nesameletus ornatus*; *Olinga feredayi*; *Triplectides obsoleta*; *Zephlebia dentata*]

Hudson, GV, 'Evidence of memory and reasoning in a pompilid', *The Entomologist's Monthly Magazine*, no. 50, 1914, p. 121. [*Sphictostethus fugax*]

Hudson, GV, *New Zealand Beetles and their Larvae*, Fergusson & Osborn Ltd, Wellington, 1934. [*Neocicindela tuberculata*; *Pericoptus truncatus*]

Hughes, RR and LT Woolcock, '*Aphaereta aotea* sp. n. (Hymenoptera: Braconidae), an alysiine parasite of dung breeding flies', *Journal of the Australian Entomological Society*, vol. 15, 1975, pp. 191–196. [*Aphaereta aotea*]

Hunt, R, 'Batfly', iss. 81, 2006, nzgeo.com/stories/batfly/. [*Mystacinobia zelandica*]

Kavanagh, ME, 'The efficiency of sound production in two cricket species, *Gryllotalpa australis* and *Teleogryllus commodus* (Orthoptera: Grylloidea)', *Journal of Experimental Biology*, vol. 130, 1987, pp. 107–119. [*Teleogryllus commodus*]

Larivière, M-C, 'Cydnidae, Acanthosomatidae, and Pentatomidae (Insecta: Heteroptera): systematics, geographical distribution, and bioecology', *Fauna of New Zealand*, no. 35, Manaaki Whenua Press, Lincoln, 1995, pp. 1–107. [*Cermatulus nasalis*]

Larivière, M-C, D Burckhardt and A Larochelle, 'Peloridiidae (Insecta: Hemiptera: Coleorrhyncha)', *Fauna of New Zealand*, no. 67, Manaaki Whenua Press, Lincoln, 2011, pp. 1–78. [Family Peloridiidae]

Larivière, M-C, MJ Fletcher and A Larochelle, 'Auchenorrhyncha (Insecta: Hemiptera): catalogue', *Fauna of New Zealand*, no. 63, Manaaki Whenua Press, Lincoln, 2010, pp. 1–232. [*Kikihia* spp.]

Larivière, M-C and A Larochelle, 'Heteroptera (Insecta: Hemiptera): catalogue', *Fauna of New Zealand*, no. 50, Manaaki Whenua Press, Lincoln, 2004, p. 330. [*Hydrometra strigosa*]

Larochelle, A and M-C Larivière, 'Carabidae (Insecta: Coleoptera): synopsis of species, Cicindelinae to Trechinae (in part)', *Fauna of New Zealand*, no. 69, Manaaki Whenua Press, Lincoln, 2013, pp. 1–193. [*Neocicindela tuberculata*]

Logan, DP, 'Nymphal development and lifecycle length of *Kikihia ochrina* (Walker) (Homoptera: Cicadidae)', *The Weta*, vol. 31, 2006, pp. 19–22. [*Kikihia* spp.]

Macfarlane, RP, PA Maddison, IA Andrew, JA Berry, P Johns, RJB Hoare, M-C Larivière, P Greenslade, RC Henderson, CN Smithers, R Palma, JB Ward, RLC Pilgrim, RAB Leschen, D Towns, I McLellan, D Teulon, JF Lawrence, G Kuschel, D Burckhardt, TR Buckley and SA Trewick, 'Phylum Arthropoda. Subphylum Hexapoda. Protura, springtails, Diplura, and insects', in DP Gordon (ed.), *New Zealand Inventory of Biodiversity*, vol. 2, Canterbury University Press, Christchurch, 2010, Chap. 9, pp. 233–467. [*Argosarchus horridus*; Family Blatellidae and other cockroaches]

Marinov, M and M Ashbee, *Dragonflies and Damselflies of New Zealand*, Auckland University Press, Auckland, 2019. [*Uropetala carovei; Xanthocnemis zealandica*]

Mark-Chan, CJ, JC O'Hanlon and GI Holwell, 'Camouflage in lichen moths: Field predation experiments and avian vision modelling demonstrate the importance of wing pattern elements and background for survival', *Journal of Animal Ecology*, vol. 91, no. 12, pp. 2358–2369, doi: 10.1111/1365-2656.13817. [*Declana atronivea*]

Mason, PC, 'Alpine grasshoppers (Orthoptera: Acrididae) in the southern alps of Canterbury, New Zealand', PhD thesis, University of Canterbury, Christchurch, 1971. [*Brachaspis robustus*]

May, BM, 'Larvae of Curculionoidea (Insecta: Coleoptera): a systematic overview', *Fauna of New Zealand*, no. 28, Manaaki Whenua Press, Lincoln, 1993, pp. 1–226. [*Lasiorhynchus barbicornis*]

McIntyre, M, 'The ecology of some large wētā species in New Zealand', in LH Field (ed.), *The Biology of Wetas, King Crickets and Their Allies*, CABI Publishing, Wallingford, 2001, pp. 225–242. [*Deinacrida rugosa; Motuweta isolata*]

Meads, MJ, 'Some observations on *Lasiorhynchus barbicornis* (Brentidae: Coleoptera)', *New Zealand Entomologist*, vol. 6, no. 2, 1976, pp. 171–176. doi: 10.1080/00779962.1976.9722234. [*Lasiorhynchus barbicornis*]

Meads, MJ, *Forgotten fauna*, DSIR Publishing, Wellington, 1990, pp. 1–94. [*Exsul singularis; Mystacinobia zelandica*]

Meads, MJ, *The Weta Book – A guide to the identification of wetas*, DSIR Land Resources, Lower Hutt, 1990. [*Pachyrhamma* spp.]

Miller, D, *Common Insects in New Zealand*, revised edition, AH & AW Reed Ltd, Wellington, 1984 [*Aoraia dinodes*; Family Blatellidae and other cockroaches; *Huberia striata*; *Ophiocordyceps robertsii*; *Orthodera novaezealandiae*; *Weeleus acutus*]

Miskelly, C, 'A plague of flax weevils – a conservation hyper-success story', Museum of New Zealand Te Papa Tongarewa blog, 13 November 2013, blog. tepapa.govt.nz/2013/11/13/a-plague-of-flax-weevils-a-conservation-hyper-success-story. [*Anagotus fairburni*]

Miskelly, C, 'An inordinate fondness for weevils', Museum of New Zealand Te Papa Tongarewa blog, 15 December 2016, blog.tepapa.govt.nz/2016/12/15/an-inordinate-fondness-for-weevils. [*Anagotus fairburni*]

Mound, LA and AK Walker, 'Tubulifera (Insecta: Thysanoptera)', *Fauna of New Zealand*, no. 10, Department of Scientific and Industrial Research, Wellington, 1986, pp. 1–144. [*Cleistothrips idolothripoides*]

New Zealand Farm Forestry Association, 'Pests and diseases of forestry in New Zealand: Huhu beetle, *Prionoplus reticularis*', Wellington, revised 2009, nzffa.org.nz/farm-forestry-model/the-essentials/forest-health-pests-and-diseases/Pests/Prionoplus-reticularis/huhu-beetle. [*Prionoplus reticularis*]

New Zealand Farm Forestry Association, 'Pests and diseases of forestry in New Zealand: Lemon-tree borer', Wellington, revised 2009, nzffa.org.nz/farm-forestry-model/the-essentials/forest-health-pests-and-diseases/Pests/Oemona-hirta/lemon-tree-borer. *[Oemona hirta]*

New Zealand Farm Forestry Association, 'Pests and diseases of forestry in New Zealand: New Zealand wetwood termites', Wellington, revised 2009, nzffa.org.nz/farm-forestry-model/the-essentials/forest-health-pests-and-diseases/Pests/Stolotermes/new-zealand-wetwood-termites/. *[Stolotermes ruficeps]*

New Zealand Farm Forestry Association, 'Pests and diseases of forestry in New Zealand: Puriri moth', Wellington, revised 2009, nzffa.org.nz/farm-forestry-model/the-essentials/forest-health-pests-and-diseases/Pests/Puriri-moth/Puriri-mothEnt16/. *[Aenetus virescens]*

Ordish, RG, 'Aggregation and communication of the Wellington weta *Hemideina crassidens* (Blanchard) (Orthoptera: Stenopelmatidae)', *New Zealand Entomologist*, vol. 15, 1992, pp. 1–8. *[Hemideina crassidens]*

Palma, RL, 'Phthiraptera (Insecta): A catalogue of parasitic lice from New Zealand', *Fauna of New Zealand*, no. 76, Landcare Research, Lincoln, 2007, pp. 1–400. **[Family Philopteridae]**

Patrick, BH, 'The status of the bat-winged fly, *Exsul singularis* Hutton (Diptera: Muscidae: Coenosiinae)', *New Zealand Entomologist*, vol. 19, no. 1, 1996, doi: 10.1080/00779962.1996.9722018. *[Exsul singularis]*

Ratcliffe, B and J Orozco, 'A review of the biology of *Pericoptus truncatus* (Fabr.) (Coleoptera: Scarabaeidae: Pentodontini) from New Zealand and a revised description of the third instar', *The Coleopterists Bulletin*, vol. 63, 2009, pp. 445–451, doi: 10.1649/1186.1. *[Pericoptus truncatus]*

Reaney LT, JM Drayton and MD Jennions, 'The role of body size and fighting experience in predicting contest behaviour in the black field cricket, *Teleogryllus commodus*', *Behavioral Ecology and Sociobiology*, vol. 65, 2010, pp. 217–225. *[Teleogryllus commodus]*

Rowe, RJ, *The Dragonflies of New Zealand*, Auckland University Press, Auckland, 1987. *[Uropetala carovei; Xanthocnemis zealandica]*

Salmon, JT, *The Stick Insects of New Zealand*, Raupo, Auckland, 1991. *[Argosarchus horridus]*

Science Learning Hub Pokapū Akoranga Pūtaiao, 'Vegetable caterpillar', 30 April 2009, updated 21 November 2018, sciencelearn.org.nz/resources/1435-vegetable-caterpillar. *[Aoraia dinodes; Ophiocordyceps robertsii]*

Sharell, R, *New Zealand Insects and Their Story*, revised edition, William Collins Publishers Ltd, Auckland, 1982. *[Declana atronivea]*

Smit, FGA, 'The fleas of New Zealand (Siphonaptera)', *Journal of the Royal Society of New Zealand*, Vol. 9, No. 2, 1979, pp. 143–232. [*Parapsyllus* spp.]

Stringer, IAN, 'Distinguishing Mercury Islands tusked weta, *Motuweta isolata*, from a ground weta, *Hemiandrus pallitarsis* (Orthoptera: Anostostomatidae) in the field, with observations of their activity', DOC Research & Development Series 258, Department of Conservation, Wellington, 2006. [*Motuweta isolata*]

Stringer, IAN and R Chappell, 'Possible rescue from extinction: transfer of a rare New Zealand tusked weta to islands in the Mercury group', *Journal of Insect Conservation*, vol. 12, pp. 371–382. [*Motuweta isolata*]

Te Manahuna Aoraki Project, 'World first predator-exclusion fence for insects', 19 December 2019, temanahunaaoraki.org/world-first-predator-exclusion-fence-for-insects. [*Brachaspis robustus*]

University of Wisconsin at Milwaukee, College of Letters & Science Field Station, 'Marsh treader', 26 April 2011, uwm.edu/field-station/marsh-treader/. [*Hydrometra strigosa*]

Walker, A, *The Reed Handbook of Common New Zealand Insects*, Raupo Publishing, Auckland, 2000. [*Stenoperla prasina*; *Orthodera novaezealandiae*]

Ware, JL, CD Beatty, M Sanchez Herrera, S Valley, J Johnson, C Kerst, ML May, and G Theishinger, 'The petaltail dragonflies (Odonata: Petaluridae): Mesozoic habitat specialists that survive to the modern day', *Journal of Biogeography*, vol. 41, no. 7, 2014, pp. 1291–1300, doi: 10.1111/jbi.12273. [*Uropetala carovei*]

White, EG, 'Ecological research and monitoring of the protected grasshopper *Brachaspis robustus* in the Mackenzie Basin', Department of Conservation, Wellington, 1994, pp. 1–50. [*Brachaspis robustus*]

Winterbourn, MJ, KLD Gregson and CH Dolphin, 'Guide to the aquatic insects of New Zealand, fourth edition', *Bulletin of the Entomological Society of New Zealand*, vol. 14, pp. 1–108. [*Ameletopsis perscitus*; *Aoteapsyche colonica*; *Archichauliodes diversus*; *Helicopsyche zealandica*; *Nesameletus ornatus*; *Olinga feredayi*; *Triplectides obsoleta*; *Zephlebia dentata*]

Wise, KAJ, 'Distribution and zoogeography of New Zealand Megaloptera and Neuroptera', *Records of the Auckland Institute and Museum*, vol. 28, 1991, pp. 211–227. [*Archichauliodes diversus*]

ACKNOWLEDGEMENTS

We would like to acknowledge George Gibbs (a descendant of GV Hudson) for approval to use Hudson's illustrations and for his helpful comments on an initial draft of this book. Ricardo Palma is thanked for his kind permission to reproduce his illustration of a feather louse, and we would like to express our appreciation of Des Helmore's artistry and the availability of so much of his work on Wikicommons.

Many thanks to Tim Denee for the cover and series design, Sarah Elworthy for the typesetting, Teresa McIntyre for the copy edit, and Mike Wagg and Caren Wilton for the proof-reads.

IMAGE CREDITS

The illustrations on pages 14, 38, 42, 44, 46, 48, 50, 52, 54, 56, 58, 60, 64, 66, 68 (modified), 76, 78, 80, 82, 84, 86, 88, 94, 96, 98, 108, 110 (modified), 112, 116 and 118 are by Des Helmore / Manaaki Whenua – Landcare Research from Wikimedia Commons, commons.wikimedia.org

The illustrations on pages 20, 22, 24, 26, 28, 30, 32, 34, 36, 40, 70, 72, 74, 90, 92, 100, 102, 104, 106 and 114 are from hand painted plates by George Vernon Hudson and reproduced with the permission of his family © Museum of New Zealand Te Papa Tongarewa.

The illustration on page 62 is by Ricardo Palma in 'A revision of the genus *Naubates* (Insecta: Phthiraptera: Philopteridae), *Journal of the Royal Society of New Zealand*, vol 32, issue 11, 2010. Reproduced with the permission of the publisher, Taylor & Francis tandfonline.com

INDEX OF SPECIES

Bold page numbers refer to species descriptions.

ABOUT THE AUTHORS

Dr Julia Kasper is Lead Curator Invertebrates at Te Papa and an entomologist specialising in flies. She researches lower Diptera in Aotearoa New Zealand with a strong focus on biosecurity and conservation, e.g. invasive mosquito species, Diptera species in fresh water in and around cave systems, and pollination networks.

Julia is an advisor for the Ministry of Health when it comes to blood-sucking and disease-transmitting flies, and is New Zealand's Forensic Entomologist, assisting the police in legal-medical cases that involve insect evidence. She also has a strong background in science communication and is an advocate for insects, keen to improve their perception in the public eye.

Dr Phil Sirvid is Assistant Curator in the Natural History Team at Te Papa. Phil has a broad general knowledge of Aotearoa New Zealand entomology but specialises in arachnids, particularly spiders and harvestmen. He has published on the taxonomy, systematics and evolutionary history of New Zealand spiders as well as on medical entomology and arachnid conservation.

Phil is the co-author of *Why is That Spider Dancing? The Amazing Arachnids of Aotearoa*, which was a finalist for the New Zealand Book Awards for Children and Young Adults in 2022.

First published in New Zealand in 2023 by
Te Papa Press, PO Box 467, Wellington, New Zealand
www.tepapapress.co.nz

Text: Julia Kasper and Phil Sirvid
© Museum of New Zealand Te Papa Tongarewa
Images: as credited on page 131.

TE PAPA® is the trademark of the Museum of
New Zealand Te Papa Tongarewa
Te Papa Press is an imprint of the Museum of
New Zealand Te Papa Tongarewa

A catalogue record is available from the National Library
of New Zealand

978-1-99-116554-1

Cover and internal design by Tim Denee
Cover illustration based on the kapowai or giant bush
dragonfly (*Uropetala carovei*)
Typesetting by Sarah Elworthy
Digital imaging by Yoan Jolly

Printed by Everbest Printing Investment Limited